A Complete Book on Nature of Science and Biology

A Complete Book on

Nature of Science and Biology

C.S. Chandoliya

CYBER TECH PUBLICATIONS

4264/3, Ansari Road, Daryaganj, New Delhi-110002 (India)

Ph.: 011-23244078. Fax: 011-23280028

E-Mail: cyberpublicationsdelhi@yahoo.com

A COMPLETE BOOK ON NATURE OF SCIENCE AND BIOLOGY

C.S. Chandoliya

First Edition 2009

Published by :

G.S. Rawat for **Cyber Tech Publications**
4264/3, Ansari Road, Daryaganj, New Delhi-110002 (India)
Ph.: 011-23244078. Fax: 011-23280028
E-Mail: cyberpublicationsdelhi@yahoo.com

Printed at Asian Offset Printers

Preface

We enter a new century, one that is exploding with excitement in biology. This first year of the new millennium has seen the completion of the Human Genome Project, with the full sequence of the human genome now available for research and exploration. Embryonic stem cells were cloned for the first time in the year 2000, and offer the potential for curing a wide range of ills, from spinal cord injuries to diabetes. Golden rice, a genetically modified crop to which has been added a battery of genes that overcome deficiencies in vitamin A and iron, was planted for the first time in Asian fields. Neurobiologists for the first time caught a glimpse of the molecular basis of learning. Even taxonomy, that bastion of conservative judgments, seems to be undergoing a sea change, with molecular phylogenies forcing the redrawing of many family trees, from angiosperm plants to insects and other arthropods.

In the last few decades, the pace of biological research has accelerated, as we have learned for the first time how to manipulate genes. In agriculture this has led to waves of controversy, in medicine to advances universally applauded. But no matter how one views genetic engineering, no one questions that it is changing the science of biology in profound ways.

What is important about these changes in biology, what excites us like no past year, is the potential to influence our health, and that of our world. Biology as a science can --indeed, must --be more than simply a trip to the zoo, an investigation of what living things are like and how they work. These things are important parts of biology, of course, the knowledge that provides the core of the science. But it can't stop there. The knowledge of biology that has been gained, especially in the last decade, provides us with a tool of unprecedented power to improve the human condition and lessen human impact on the world we share with life's other creatures.

—Author

CONTENTS

1
INTRODUCTION

After all, many questions in biology can be answered without mentioning evolution: How do birds fly? How can certain plants grow in the desert? Why do children resemble their parents? Each of these questions has an immediate answer involving aerodynamics, the storage and use of water by plants, or the mechanisms of heredity.

Students ask about such things all the time. The answers to these questions often raise deeper questions that are sometimes asked by students: How did things come to be that way? What is the advantage to birds of flying? How did desert plants come to differ from others? How did an individual organism come to have its particular genetic endowment? Answering questions like these requires a historical context—a framework of understanding that recognizes change through time.

People who study nature closely have always asked these kinds of questions. Over time, two observations have proved to be especially perplexing. The older of these has to do with the diversity of life: Why are there so many different kinds of plants and animals? The more we explore the world, the more impressed we are with the multiplicity of kinds of organisms. In the mid-nineteenth century, when Charles Darwin was writing *On the Origin of Species*, naturalists recognized several tens of thousands of different plant and animal species. By the middle of the twentieth century, biologists had paid more attention to less conspicuous forms of life, from insects to microorganisms, and the estimate was up to 1 or 2 million. Since then, investigations in tropical rain forests—the center of much of the world's biological diversity—have multiplied those estimates at least tenfold. What process has created this extraordinary variety of life?

The second question involves the inverse of life's diversity. How can the similarities among organisms be explained? Humans have always noticed the similarities among closely related species, but it gradually became apparent that even distantly related species share many anatomical and functional characteristics. The bones in a whale's front flippers are arranged in much the same way as the bones in our own arms. As organisms grow from fertilized egg cells into embryos, they pass through many similar developmental stages. Furthermore, as paleontologists studied the fossil record, they discovered countless extinct species that are clearly related in various ways to organisms living today.

This question has emerged with even greater force as modern experimental biology has focused on processes at the cellular and molecular level. From bacteria to yeast to mice to humans, all living things use the same biochemical machinery to carry out the basic processes of life. Many of the proteins that make up cells and catalyze chemical reactions in the body are virtually identical across species. Certain human genes that code for proteins differ little from the corresponding genes in fruit flies, mice, and primates. All living things use the same biochemical system to pass genetic information from one generation to another.

From a scientific standpoint, there is one compelling answer to questions about life's commonalities. Different kinds of organisms share so many characteristics of structure and function because they are related to one another. But how?

SOLVING THE PUZZLE

The concept of biological evolution addresses both of these fundamental questions. It accounts for the relatedness among organisms by explaining that the millions of different species of plants, animals, and microorganisms that live on earth today are related by descent from common ancestors—like distant cousins. Organisms in nature typically produce more offspring than can survive and reproduce given the constraints of food, space, and other resources in the environment. These offspring often differ from one another in ways that are heritable—that is, they can pass on the differences genetically to their own offspring. If competing offspring have traits that are advantageous in a given environment, they will survive and pass on those traits. As differences continue to accumulate over generations, populations of organisms diverge from their ancestors.

This straightforward process, which is a natural consequence of biologically reproducing organisms competing for limited resources, is responsible for one of the most magnificent chronicles known to science. Over billions of years, it has led the earliest organisms on earth to diversify into all of the plants, animals, and microorganisms that exist today. Though humans, fish, and bacteria would seem to be so different as to defy comparison, they all share some of the characteristics of their common ancestors. Evolution also explains the great diversity of modern species.

Populations of organisms with characteristics enabling them to occupy ecological niches not occupied by similar organisms have a greater chance of surviving. Over time—as the next chapter discusses in more detail—species have diversified and have occupied more and more ecological niches to take advantage of new resources. Evolution explains something else as well. During the billions of years that life has been on earth, it has played an increasingly important role in altering the planet's physical environment.

For example, the composition of our atmosphere is partly a consequence of living systems. During photosynthesis, which is a product of evolution, green plants absorb carbon dioxide and water, produce organic compounds, and release oxygen. This process has created and continues to maintain an atmosphere rich in oxygen. Living communities also profoundly affect weather and the movement of water among the oceans, atmosphere, and land. Much of the rainfall in the forests of the western Amazon basin consists of water that has already made one or more recent trips through a living plant. In addition, plants and soil microorganisms exert important controls over global temperature by absorbing or emitting "greenhouse gases" (such as carbon dioxide and methane) that increase the earth's capacity to retain heat. In short, biological evolution accounts for three of the most fundamental features of the world around us: the similarities among living things, the diversity of life, and many features of the physical world we inhabit. Explanations of these phenomena in terms of evolution draw on results from physics, chemistry, geology, many areas of biology, and other sciences. Thus, evolution is the central organizing principle that biologists use to understand the world. To teach biology without explaining evolution deprives students of a powerful concept that brings great order and coherence to our understanding of life.

The teaching of evolution also has great practical value for students. Directly or indirectly, evolutionary biology has made many contributions to society. Evolution explains why many human pathogens have been developing resistance to formerly effective drugs and suggests ways of confronting this increasingly serious problem (this issue is discussed in greater detail in Chapter 2). Evolutionary biology has also contributed to many important agricultural advances by explaining the relationships among wild and domesticated plants and animals and their natural enemies. An understanding of evolution has been essential in finding and using natural resources, such as fossil fuels, and it will be indispensable as human societies strive to establish sustainable relationships with the natural environment.

Such examples can be multiplied many times. Evolutionary research is one of the most active fields of biology today, and discoveries with important practical applications occur on a regular basis.

Those who oppose the teaching of evolution in public schools sometimes ask that teachers present "the evidence against evolution." However, there is no debate within the scientific community over whether evolution occurred, and there is no evidence that evolution has not occurred. Some of the details of how evolution occurs are still being investigated. But scientists continue to debate only the particular mechanisms that result in evolution, not the overall accuracy of evolution as the explanation of life's history.

EVOLUTION AND THE NATURE OF SCIENCE

Teaching about evolution has another important function. Because some people see evolution as conflicting with widely held beliefs, the teaching of evolution offers educators a superb opportunity to illuminate the nature of science and to differentiate science from other forms of human endeavor and understanding.

Chapter 3 describes the nature of science in detail. However, it is important from the outset to understand how the meanings of certain key words in science differ from the way that those words are used in everyday life.

Think, for example, of how people usually use the word "theory." Someone might refer to an idea and then add, "But that's only a theory." Or someone might preface a remark by saying, "My theory is...." In common usage, theory often means "guess" or "hunch." In science, the word "theory" means something quite different. It refers to an overarching explanation that has been well substantiated. Science has many other powerful theories besides evolution. Cell theory says that all living things are composed of cells. The heliocentric theory says that the earth revolves around the sun rather than vice versa. Such concepts are supported by such abundant observational and experimental evidence that they are no longer questioned in science.

Sometimes scientists themselves use the word "theory" loosely and apply it to tentative explanations that lack well-established evidence. But it is important to distinguish these casual uses of the word "theory" with its use to describe concepts such as evolution that are supported by overwhelming evidence. Scientists might wish that they had a word other than "theory" to apply to such enduring explanations of the natural world, but the term is too deeply engrained in science to be discarded.

As with all scientific knowledge, a theory can be refined or even replaced by an alternative theory in light of new and compelling evidence. For example, Chapter 3 describes how the geocentric theory that the sun revolves around the earth was replaced by the heliocentric theory of the earth's rotation on its axis and revolution around the sun. However, ideas are not referred to as "theories" in science unless they are supported by bodies of evidence that make their subsequent abandonment very unlikely. When a theory is supported by as much evidence as evolution, it is held with a very high degree of confidence.

In science, the word "hypothesis" conveys the tentativeness inherent in the common use of the word "theory." A hypothesis is a testable statement about the natural world. Through experiment and observation, hypotheses can be supported or rejected. As the earliest level of understanding, hypotheses can be used to construct more complex inferences and explanations. Like "theory," the word "fact" has a different meaning in science than it does in common usage. A scientific fact is an observation that has been confirmed over and over. However, observations are gathered by our senses, which can never be trusted entirely. Observations also can change with better technologies or with better ways of looking at data. For example, it was held as a scientific fact for many years that human cells have 24 pairs of chromosomes, until improved techniques of microscopy revealed that they actually have 23. Ironically, facts in science often are more susceptible to change than theories— which is one reason why the word "fact" is not much used in science.

Finally, "laws" in science are typically descriptions of how the physical world behaves under certain circumstances. For example, the laws of motion describe how objects move when subjected to certain forces. These laws can be very useful in supporting hypotheses and theories, but like all elements of science they can be altered with new information and observations.

Those who oppose the teaching of evolution often say that evolution should be taught as a "theory, not as a fact." This statement confuses the common use of these words with the scientific use. In science, theories do not turn into facts through the accumulation of evidence. Rather, theories are the end points of science. They are understandings that develop from extensive observation, experimentation, and creative reflection. They incorporate a large body of scientific facts, laws, tested hypotheses, and logical inferences. In this sense, evolution is one of the strongest and most useful scientific theories we have.

EVOLUTION AND EVERYDAY LIFE

The concept of evolution has an importance in education that goes beyond its power as a scientific explanation. All of us live in a world where the pace of change is accelerating. Today's children will face more new experiences and different conditions than their parents or teachers have had to face in their lives.

The story of evolution is one chapter— perhaps the most important one—in a scientific revolution that has occupied much of the past four centuries. The central feature of this revolution has been the abandonment of one notion about stability after another: that the earth was the center of the universe, that the world's living things are unchangeable, that the continents of the earth are held rigidly in place, and so on. Fluidity and change have become central to our understanding of the world around us. To accept the probability of change—and to see change as an agent of opportunity rather than as a threat—is a silent message and challenge in the lesson of evolution. The following dialogue dramatizes some of the problems educators encounter in teaching evolution and demonstrates ways of overcoming these obstacles.

THE CHALLENGE TO TEACHERS

Teaching evolution presents special challenges to science teachers. Sources of support upon which teachers can draw include highquality curricula, adequate preparation, exposure to information useful in documenting the evidence for evolution, and resources and contacts provided by professional associations.

One important source of support for teachers is to share problems and explore solutions with other teachers. The following vignette illustrates how a group of teachers— in this case, three biology teachers at a large public high school—can work together to solve problems and learn from each other. It is the first week of classes at Central High School. As the bell rings for third period, Karen, the newest teacher on the faculty, walks into the teachers' lounge. She greets her colleagues, Barbara and Doug. "How are your first few days going?"

asks Doug.

"Fine," Karen replies. "The secundperiod Biology I class is full, but it'll be okay. By the way, Barbara, thanks for letting me see your syllabus for Bio I. But I wanted to ask you about teaching evolution— I didn't see it there."

"You didn't see it on my syllabus because it's not a separate topic," Barbara says. "I use evolution as a theme to tie the course together, so it comes into just about every unit. You'll see a section called 'History of Life' on the second page, and there's a section called 'Natural Selection.' But I don't treat evolution separately because it is related to almost every other topic in biology."1

"Wait a minute, Barbara," Doug says. "Is that good advice for a new teacher? I mean, evolution is a controversial subject, and a lot of us just don't get around to teaching it. I don't. You do, but you're braver than most of us."

"It's not a matter of bravery, Doug," Barbara replies. "It's a matter of what needs to be taught if we want students to understand biology. Teaching biology without evolution would be like teaching civics and never mentioning the United States Constitution."

"But how can you be sure that evolution is all that important. Aren't there a lot of scientists who don't believe in evolution? Say it's too improbable?"

"The debate in science is over some of the details of how evolution occurred, not whether evolution happened or not. A lot of science and science education organizations have made statements about why it is important to teach evolution...."

"I saw a news report when I was a student,"

Karen interjects, "about a school district or state that put a disclaimer against evolution in all their biology textbooks. It said that students didn't need to believe in evolution because it wasn't a fact, only a theory. The argument was that no one really knows how life began or how it evolved because no one was there to see it happen."3 "If I taught evolution, I'd sure teach it as a theory—not a fact," says Doug.

"Just like gravity," Barbara says.

"Now, Barbara, gravity is a fact, not a theory."

"Not in scientific terms. The fact is that things fall. The explanation for why things fall is the theory of gravitation. Our problem is definitions. You're using 'fact' and 'theory' the way we use them in everyday life, but we need to use them as scientists use them. In science, a 'fact' is an observation that has been made so many times that it's assumed to be okay. How facts are explained is where theories come in: theories are explanations of what we observe. One place where students get confused about evolution is that they think of 'theory' as meaning 'guess' or 'hunch.' But evolution isn't a hunch. It's a scientific explanation, and a very good one."

"But how good a theory is it?" asks Doug. "We don't know everything about evolution."

"That's true," says Karen. "A student in one of my classes at the university told me that there are big gaps in the fossil record. Do you know anything about that?" "Well, there's *Archaeopteryx*," says Doug. "It's a fossil that has feathers like a bird but the skeleton of a small dinosaur. It's one of those missing links that's not missing any more."

"In fact, there are good transitional fossils between primitive fish and amphibians and between reptiles and mammals,"

Barbara says. "Our knowledge of fossil intermediates is actually pretty good.4 And, Doug, it sounds like you know more about evolution than you're letting on. Why don't you teach it?"

"I don't want any trouble. Every time I teach evolution, I have a student announce that 'evolution is against his religion.'" "But most of the major religious denominations have taken official positions that accept evolution," says Barbara. "One semester a friend of mine in the middle school started out her Life Science unit by having her students interview their ministers or priests or rabbis about their religion's views on evolution.

She said that most of her students came back really surprised. 'Hey,' they said, 'evolution is okay.' It defused the controversy in her class." "She didn't have Stanley in her class,"

says Doug.

"Who's Stanley?" asks Karen.

"The son of a school board member.

Given his family's religious views, I'm sure he would not come back saying evolution was okay."

"That can be a hard situation," says Barbara. "But even if Stanley came back to class saying that his religion does not accept evolution, it could help a teacher show that there are many different religious views about evolution. That's the point: religious people can still accept evolution."

"Stanley will never believe in evolution."

"We talk about 'believing' in evolution, but that's not necessarily the right word. We accept evolution as the best scientific explanation for a lot of observations—about fossils and biochemistry and evolutionary changes we can actually see, like how bacteria become resistant to certain medicines. That's why people accepted the idea that the earth goes around the sun—because it accounted for many different observations that we make. In science, when a better explanation comes around, it replaces earlier ones."

"Does that mean that evolution will be replaced by a better theory some day?" asks Karen.

"It's not likely. Not all old theories are replaced, and evolution has been tested and has a lot of evidence to support it. The point is that doing science requires being willing to refine our theories to be consistent with new information."

"But there's still Stanley," says Doug. "He doesn't even want to hear about evolution."

"I had Stanley's sister in AP biology one year," Barbara replies. "She raised a fuss about evolution, and I told her that I wasn't going to grade her on her opinion of evolution but on her knowledge of the facts and concepts. She seemed satisfied with that and actually got an A in the class."

"I still think that if you teach evolution, it's only fair to teach both."

"What do you mean by both?" asks Barbara. "If you mean both evolution and creationism, what kind of creationism do you want to teach? Will you teach evolution and the Bible? What about other religions like Buddhism or the views of Native Americans? It's hard to argue for 'both' when there are a whole lot more than two options."

"I can't teach a whole bunch of creation stories in my Bio class," says Doug. "That's the point. We can't add subjects to the science curriculum to be fair to groups that hold certain beliefs. Teaching ecology isn't fair to the polluter, either. Biology is a science class, and what should be taught is science."

"But isn't there something called 'creation science'?" asks Karen. "Can creationism be made scientific?"

"That's an interesting story. 'Creation science' is the idea that scientific evidence can support a literal interpretation of Genesis—that the whole universe was created all at once about 10,000 years ago."

"It doesn't sound very likely."

"It's not. Scientists have looked at the arguments and have found they are not supported by verifiable data. Still, back in the early 1980s, some states passed laws requiring that 'creation science' be taught whenever evolution was taught. But the Supreme Court threw out 'equal time' laws, saying that because creationism was inherently a religious and not a scientific idea, it couldn't be presented as 'truth' in science classes in the public schools."5

"Well, I'm willing to teach evolution,"

says Karen, "and I'd like to try it your way, Barbara, as a theme that ties biology together. But I really don't know enough about evolution to do it. Do you have any suggestions about where I can get information?"

"Sure, I'd be glad to share what I have. But an important part of teaching evolution has to do with explaining the nature of science. I'm trying out a demonstration after school today that I'm going to use with my Bio I class tomorrow. Why don't you both come by and we can try it out?"

"Okay," say Karen and Doug. "We'll see you then."

MAJOR THEMES IN EVOLUTION

The world around us changes. This simple fact is obvious everywhere we look. Streams wash dirt and stones from higher places to lower places. Untended gardens fill with weeds.

Other changes are more gradual but much more dramatic when viewed over long time scales. Powerful telescopes reveal new stars coalescing from galactic dust, just as our sun did more than 4.5 billion years ago. The earth itself formed shortly thereafter, when rock, dust, and gas circling the sun condensed into the planets of our solar system. Fossils of primitive microorganisms show that life had emerged on earth by about 3.8 billion years ago. Similarly, the fossil record reveals profound changes in the kinds of living things that have inhabited our planet over its long history. Trilobites that populated the seas hundreds of millions of years ago no longer crawl about. Mammals now live in a world that was once dominated by reptilian giants such as *Tyrannosaurus rex*. More than 99 percent of the species that have ever lived on the earth are now extinct, either because all of the members of the species died, the species evolved into a new species, or it split into two or more new species.

Many kinds of cumulative change through time have been described by the term "evolution," and the term is used in astronomy, geology, biology, anthropology, and other sciences.

This document focuses on the changes in living things during the long history of life on earth—on what is called biological evolution.

The ancient Greeks were already speculating about the origins of life and changes in species over time. More than 2,500 years ago, the Greek philosopher Anaximander thought that a gradual evolution had created the world's organic coherence from a formless condition, and he had a fairly modern view of the transformation of aquatic species into terrestrial ones. Following the rise of Christianity, Westerners generally accepted the explanation provided in Genesis, the first book of the Judeo-Christian-Muslim Bible, that God created everything in its present form over the course of six days. However, other explanations existed even then. Among Christian theologians, for example, Saint Thomas Aquinas (1225 to 1274) stated that the earth had received the power to produce organisms and criticized the idea that species had originated in accordance with the timetables in Genesis.1

During the early 1800s, many naturalists speculated about changes in organisms, especially as geological investigations revealed the rich story laid out in the fossilized remains of extinct creatures. But although ideas about evolution were proposed, they never gained wide acceptance because no one was able to propose a plausible mechanism for how the form of an organism might change from one generation to another. Then, in 1858, two English naturalists—Charles Darwin and Alfred Russel Wallace—simultaneously issued papers proposing such a mechanism. Both men observed that the individual members of a particular species are not identical but can differ in many ways. For example, some will be able to run a little faster, have a different color, or respond to the same circumstance in different ways. (Humans— including any class of high school students— have many such differences.) Both men further observed that many of these differences are inherited and can be passed on to offspring. This conclusion was evident from the experiences of plant and animal breeders.

Darwin and Wallace were both deeply influenced by the realization that, even though most species produce an abundance of offspring, the size of the overall population usually remains about the same. Thus, an oak tree might produce many thousands of acorns each year, but few, if any, will survive to become full-grown trees.

Darwin—who conceived of his ideas in the 1830s but did not publish them until Wallace came to similar conclusions— presented the case for evolution in detail in his 1859 book *On the Origin of Species by Natural Selection*. Darwin proposed that there will be differences between offspring that survive and reproduce and those that do not. In particular, individuals that have heritable characteristics making them more likely to survive and reproduce in their particular environment will, on average, have a better chance of passing those characteristics on to their own offspring. In this way, as many generations pass, nature would select those individuals best suited to particular environments, a process Darwin called natural selection. Over very long times, Darwin argued, natural selection acting on varying individuals within a population of organisms could account for all of the great variety of organisms we see today, as well as for the species found as fossils.

If the central requirement of natural selection is variation within populations, what is the ultimate source of this variation? This problem plagued Darwin, and he never found the answer, although he proposed some hypotheses. Darwin did not know that a contemporary, Gregor Mendel, had provided an important part of the solution. In his classic 1865 paper describing crossbreeding of varieties of peas, Mendel demonstrated that organisms acquire traits through discrete units of heredity which later came to be known as genes. The variation produced through these inherited traits is the raw material on which natural selection acts. Mendel's paper was all but forgotten until 1890, when it was rediscovered and contributed to a growing wave of interest and research in genetics. But it was not immediately clear how to reconcile new findings about the mechanisms of inheritance with evolution through natural selection. Then, in the 1930s, a group of biologists demonstrated how the results of genetics research could both buttress and extend evolutionary theory. They showed that all variations, both slight and dramatic, arose through changes, or mutations, in genes. If a mutation enabled an organism to survive or reproduce more effectively, that mutation would tend to be preserved and spread in a population through natural selection. Evolution was thus seen to depend both on genetic mutations and on natural selection. Mutations provided abundant genetic variation, and natural selection sorted out the useful changes from the deleterious ones.

Selection by natural processes of favored variants explained many observations on the geography of species differences—why, for example, members of the same bird species might be larger and darker in the northern part of their range, and smaller and paler in the southern part. In this case, differences might be explained by the advantages of large size and dark coloration in forested, cold regions. And, if the species occupied the entire range continuously, genes favoring light color and small size would be able to flow into the northern population, and vice versa—prohibiting their separation into distinct species that are reproductively isolated from one another.

How new species are formed was a mystery that eluded biologists until information about genetics and the geographical distribution of animals and plants could be put together. As a result, it became clear that the most important source of new species is the process of geographical isolation— through which barriers to gene flow can be created. In the earlier example, the interposition of a major mountain barrier, or the origin of an intermediate desert, might create the needed isolation.

Other situations also encourage the formation of new species. Consider fish in a river that, over time, changes course so as to isolate a tributary. Or think of a set of oceanic islands, distant from the mainland and just far enough from one another that interchange among their populations is rare. These are ideal circumstances for creating reproductive barriers and allowing populations of the same species to diverge from one another under the influence of natural selection. After a time, the species become sufficiently different that even when reunited they remain reproductively isolated.

They have become so different that they are unable to interbreed.

In the 1950s, the study of evolution entered a new phase. Biologists began to be able to determine the exact molecular structure of the proteins in living things— that is, the actual sequences of the amino acids that make up each protein. Almost immediately, it became clear that certain proteins that serve the same function in different species have very similar amino acid sequences. The protein evidence was completely consistent with the idea of a common evolutionary history for the planet's living things. Even more important, this knowledge provided important clues about the history of evolution that could not be obtained through the fossil record. The discovery of the structure of DNA by Francis Crick and James Watson in 1953 extended the study of evolution to the most fundamental level. The sequence of the chemical bases in DNA both specifies the order of amino acids in proteins and determines which proteins are synthesized in which cells. In this way, DNA is the ultimate source of both change and continuity in evolution. The modification of DNA through occasional changes or rearrangements in the base sequences underlies the emergence of new traits, and thus of new species, in evolution. At the same time, all organisms use the same molecular codes to translate DNA base sequences into protein amino acid sequences. This uniformity in the genetic code is powerful evidence for the interrelatedness of living things, suggesting that all organisms presently alive share a common ancestor that can be traced back to the origins of life on earth.

One common misconception among students is that individual organisms change their characteristics in response to the environment. In other words, students often think that the environment acts on individual organisms to generate physical characteristics that can then be passed on genetically to offspring. But selection can work only on the genetic variation that already is present in any new generation, and genetic variation occurs randomly, not in response to the needs of a population or organism. In this sense, as Francois Jacob has written, evolution is a "tinkerer, not an engineer." Evolution does not design new organisms; rather, new organisms emerge from the inherent genetic variation that occurs in organisms.

Discovery of a Missing Link

> *As a zoologist I have discovered many phenomena that can be rationally explained only as products of evolution, but none so striking as the ancestor of the ants. Prior to 1967 the fossil record had yielded no specimens of wasps or other* Hymenopterous *insects that might be interpreted as the ancestors of the ants. This hypothetical form was a missing link of major importance in the study of evolution. We did have many fossils of ants dating back 50 million years. These were different species from those existing today, but their bodies still possessed the basic body form of modern ants. The missing link of ant evolution was often cited by creationists as evidence against evolution. Other ant specialists and I were convinced that the linking fossils would be found, and that most likely they would be associated with the late Mesozoic era, a time when many dinosaur and other vertebrate bones were fossilized but few insects. And that is exactly what happened. In 1967 I had the pleasure of studying two specimens collected in amber (fossilized resin) from New Jersey, and dating to the late Mesozoic about 90 million years ago. They were nearly exact intermediates between solitary wasps and the highly social modern ants, and so I gave them the scientific name* Sphecomyrma, *meaning "wasp ant." Since that time many more* Sphecomyrma *specimens of similar age have been found in the United States, Canada, and Siberia, but none belonging to the modern type. With each passing year, such fossils and other kinds of evidence tighten our conception of the evolutionary origin of this important group of insects. —Edward O. Wilson*

Genetic variation is random, but natural selection is not. Natural selection tests the combinations of genes represented in the members of a species and allows to proliferate those that confer the greatest ability to survive and reproduce. In this sense, evolution is not the simple product of random chance. The booklet *Science and Creationism: A* View from the National Academy of *Sciences* summarizes several compelling lines of evidence that demonstrate beyond any reasonable doubt that evolution occurred as a historical process and continues today. In brief:

- Fossils found in rocks of increasing age attest to the interrelated lineage of living things, from the single-celled organisms that lived billions of years ago to *Homo sapiens.* The most recent fossils closely resemble the organisms alive today, whereas increasingly older fossils are progressively different, providing compelling evidence of change through time.
- Even a casual look at different kinds of organisms reveals striking similarities among species, and anatomists have discovered that these similarities are more than skin deep. All vertebrates, for example, from fish to humans, have a common body plan characterized by a segmented body and a hollow main nerve cord along the back. The best available scientific explanation for these common structures is that all vertebrates are descended from a common ancestor species and that they have diverged through evolution.
- In the past, evolutionary relationships could be studied only by examining the consequences of genetic information, such as the anatomy, physiology, and embryology of living organisms. But the advent of molecular biology has made it possible to read the history of evolution that is written in every organism's DNA. This information has allowed organisms to be placed into a common evolutionary family tree in a much more detailed way than possible from previous evidence.

Evolution is the only plausible scientific explanation that accounts for the extensive array of observations summarized above. The concept of evolution through random genetic variation and natural selection makes sense of what would otherwise be a huge body of unconnected observations. It is no longer possible to sustain scientifically the view that the living things we see today did not evolve from earlier forms or that the human species was not produced by the same evolutionary mechanisms that apply to the rest of the living world.

The following two sections of this chapter examine two important themes in evolutionary theory. The first concerns the occurrence of evolution in "real time"— how changes come about and result in new kinds of species. The second is the ecological framework that underlies evolution, which is needed to understand the expansion of biological diversity.

EVOLUTION AS A CONTEMPORARY PROCESS

Evolution by natural selection is not only a historical process— it still operates today. For example, the continual evolution of human pathogens has come to pose one of the most serious public health problems now facing human societies. Many strains of bacteria have become increasingly resistant to once-effective antibiotics as natural selection has amplified resistant strains that arose through naturally occurring genetic variation. The microorganisms that cause malaria, gonorrhea, tuberculosis, and many other diseases have demonstrated greatly increased resistance to the antibiotics and other drugs used to treat them in the past. The continued use and overuse of antibiotics has had the effect of selecting for resistant populations because the antibiotics give these strains an advantage over nonresistant strains.

Similar episodes of rapid evolution are occurring in many different organisms. Rats have developed resistance to the poison warfarin. Many hundreds of insect species and other agricultural pests have evolved resistance to the pesticides used to combat them—and even to chemical defenses genetically engineered into plants. Species of plants have evolved tolerance to toxic metals and have reduced their interbreeding with nearby nontolerant plants—an initial step in the formation of separate species. New species of plants have arisen through the crossbreeding of native plants with plants introduced from elsewhere in the world.

The creation of a new species from a preexisting species generally requires thousands of years, so over a lifetime a single human usually can witness only a tiny part of the speciation process. Yet even that glimpse of evolution at work powerfully confirms our ideas about the history and mechanisms of evolution. For example, many closely related species have been identified that split from a common ancestor very recently in evolutionary terms. An example is provided by the North American lacewings *Chrysoperla carnea* and *Chrysoperla downesi*. The former lives in deciduous woodlands and is pale green in summer and brown in winter. The latter lives among evergreen conifers and is dark green all year round. The two species are genetically and morphologically very similar.

Yet they occupy different habitats and breed at different times of the year and so are reproductively isolated from each other. The fossil record also sheds light on speciation. A particularly dramatic example comes from recently discovered fossil evidence documenting the evolution of whales and dolphins. The fossil record shows that these cetaceans evolved from a primitive group of hoofed mammals called *Mesonychids*. Some of these mammals crushed and ate turtles, as evidenced by the shape of their teeth. This mammal gave rise to a species with front forelimbs and powerful hind legs with large feet that were adapted for paddling. This animal, known as *Ambulocetus*, could have moved between sea and land.

Its fossilized vertebrae also show that this animal could move its back in a strong up and down motion, which is the method modern cetaceans use to swim and dive. A later fossil in the series from Pakistan shows an animal with smaller functional hind limbs and even greater back flexibility. This species, *Rodhocetus*, probably did not venture onto land very often, if at all. Finally, *Basilosaurus* fossils from Egypt and the United States present a recognizable whale, with front flippers for steering and a completely flexible backbone. But this animal still has hind limbs (thought to have been nonfunctional), which have become further reduced in modern whales.

Another focus of research has been the evolution of ancient apelike creatures through many intermediate forms into modern humans. *Homo sapiens,* one of 185 known living species in the primate order, is a member of the hominoids, a category that includes orangutans, gorillas, and chimpanzees. The succession of species that would give rise to humans seems to have separated from the succession that would lead to the apes about 5 to 8 million years ago. The first members of our genus, *Homo,* had evolved by about 1.5 million years ago. According to recent evidence— based on the sequencing of DNA found in a part of human cells known as mitochondria— it has been proposed that a small group of modern humans evolved in Africa about 150,000 years ago and spread throughout the world, replacing archaic populations of *Homo sapiens.*

Ongoing Evolution Among Darwin's Finches

A particularly interesting example of contemporary evolution involves the 13 species of finches studied by Darwin on the Galapagos Islands, now known as Darwin's finches. A research group led by Peter and Rosemary Grant of Princeton University has shown that a single year of drought on the islands can drive evolutionary changes in the finches.6 Drought diminishes supplies of easily cracked nuts but permits the survival of plants that produce larger, tougher nuts. Drought thus favors birds with strong, wide beaks that can break these tougher seeds, producing populations of birds with these traits. The Grants have estimated that if droughts occur about once every 10 years on the islands, a new species of finch might arise in only about 200 years.7

EVOLUTION AND ECOLOGY

Animals and plants do not live in isolation, nor do they evolve in isolation. Indeed, much of the pressure toward diversification comes not only from physical factors in the environment but from the presence of other species. Any animal is a potential host for parasites or prey for a carnivore. A plant has other plants as competitors for space and light, can be a host for parasites, and provides food for herbivores. The interactions within the complex communities, or ecosystems, in which organisms live can generate powerful evolutionary forces.

Evolution in natural communities arises from both constraints and opportunities. The constraints come from competitors, primarily among the same species. There are only so many nest holes for bluebirds and so much food for mice. Genetically different individuals that are able to move to a different resource—a new food supply, for example, or a hitherto uninhabited area— are able to exploit that resource free of competition. As a result, the trait that opened up the new opportunity will be favored by natural selection because the individuals possessing it are able to survive and reproduce better than other members of their species in the new environment. An ecologist would say that the variant had occupied a new niche—a term that defines the "job description" of an organism.

(For example, a bluebird would have the niche of insectand fruiteater, inhabitant of forest edges and meadows, tree-hole nester, and so on.) One often finds closely related species in the same place and occupying what look like identical niches. However, if the niches were truly identical, one of the species should have a competitive advantage over the other and eventually drive the less fit species to extinction or to a different niche. That leads to a tentative hypothesis: where we find such a situation, careful observation should reveal subtle niche specialization of the apparently competing species. This hypothesis has been tested by many biologists. For example, in the 1960s Robert MacArthur carefully studied three North American warblers of the same genus that were regularly seen feeding on insects in coniferous trees in the same areas— indeed, often in the same trees. MacArthur's painstaking observations revealed that the three were actually specialists: one fed on insects on the major branches near the trunk; another occupied the mid-regions of branches and ate from different parts of the foliage; and the third fed on insects occupying the finest needles near the periphery of the tree. Although the three warblers occurred together, they were in fact not competitors for the same food resources.

Often, species that are evolving together in the same ecosystem do so through a highly interactive process. For example, natural selection will favor organisms with defenses against predation; in turn, predators experience selection for traits that overcome those defenses. Such coevolutionary competitions are common in nature. Many plants manufacture and store chemicals that deter herbivorous insects; but usually one or more insect species will have evolved biochemical mechanisms for inactivating the deterrent, providing them with a plant they can eat relatively free of competitors.

Another classic example of coevolution involves the introduction of rabbits and the myxomatosis virus into Australia. After rabbits were brought to Australia, they multiplied rapidly and threatened the wool industry because they grazed on the same plants as sheep. To control the rabbit population, a virulent pathogen of rabbits, the myxomatosis virus, also was introduced into Australia. Within a decade, rabbits had become more resistant to the virus, and the virus had evolved into a less virulent form, allowing both the host and pathogen to coexist.

CONCLUSION

As the examples in this chapter demonstrate, evolutionary biology provides an extremely active and rich source of new insights into the world. By exploring the history of life on earth and shedding light on how evolution works, evolutionary biology is linking fundamental scientific research to knowledge needed to meet important societal needs, including the preservation of our environment. Few other ideas in science have had such a far-reaching impact on our thinking about ourselves and how we relate to the world.

A Chemical Distress Signal

J. H. Tumlinson and colleagues at the U.S. Department of Agriculture's Research Service Laboratories in Gainesville, Florida, have explored a fascinating case that illustrates the intricacy of many ecological relationships. Cotton plants, like many other crops, are attacked by caterpillars. One destructive cotton pest, the army worm, produces a complex series of reactions when it feeds on the plant—a reaction that involves the caterpillar itself, the tissues of the plant, and a third participant, a wasp that preys on the caterpillar. When the caterpillar chews on the cotton plant leaf, a reaction occurs that causes the plant to synthesize and release a class of volatile chemicals that escape into the air and travel rapidly downwind. The chemicals are detected by wasps, who follow the scent and are able to find the caterpillars and deposit eggs within them. The eggs hatch, and the wasp larvae destroy the caterpillar.8 This complex case of "chemical ecology" required a series of linked coevolutionary events: the response of the plant to a special signal from its predator, and the response of the wasp to a special signal from the host of its prey.

Teaching About Evolution and the Nature of Science

In the following vignette, Barbara, Doug, and Karen use a model to continue their discussion of the nature of science and its implications for the teaching of evolution.

"Thanks for meeting with me this afternoon," Barbara says. "To begin this demonstration I first need to ask you what you think science is."

"Oh, I had that in college," says Karen. "The scientific method is to identify a question, gather information about it, develop a hypothesis that answers the question, and then do an experiment that either proves or disproves the hypothesis."

"But that was one of my points about evolution," Doug says. "No one was there when evolution happened and we can't do any experiments about what happened in the past. So by your definition, Karen, evolution isn't science."

"Science is a lot more than just supporting or rejecting hypotheses," Barbara replies. "It also involves observation, creativity, and judgment. Here's an activity I use to teach the nature of science."

Barbara takes a cardboard mailing tube about one foot long that has the ends of four ropes extending from it. As Barbara tugs on the various ropes one at a time, she has Doug and Karen make observations of what happens. After three or four pulls, she asks Karen and Doug to predict what will happen when she pulls on one of the ropes. Both are able to predict that if Barbara pulls rope A, rope B will move. Barbara then asks if there are additional manipulations they would like to see, and she follows their requests.

Barbara then asks Doug and Karen to sketch a model of what is inside the tube that could explain their observations. When Karen and Doug show their sketches to each other, they realize that they have come up with different models. Barbara asks them if they want to make any changes to their sketches based on the comparison, and both of them make modifications, although their final models are still different.

Barbara then gives them their own cardboard tubes and some string and asks them to build the model they proposed. When their models are built, Barbara holds up her tube and asks Doug and Karen to follow her actions with their own models, to see if the two models behave in the same way as Barbara's tube. But when Barbara pulls string A in her tube, Karen's model does not work the same way. Karen asks if she can make some changes in her model, and once she does her new model seems to work the same way as Barbara's. Doug's model consistently behaves the same way as Barbara's tube.

"Now wait a minute," Karen says. "What do ropes and tubes have to do with science and evolution?" "You might not know it, but what we just did is much of what science is about. You observed what happened when I pulled these ropes. Then, based on your initial observations, you made a prediction about what would happen if we manipulated the system in a specific way. How accurate was your prediction?"

"We were right," Doug responds.

"And why were you able to predict what

would happen before I pulled the rope?"

"I used what I observed in the first few pulls to help me predict what would happen later."

"Basically what each of you did was to speculate about how my tube was working on the basis of some limited observations.

Scientists do that type of thing all the time. They make observations and try to explain what's going on, or sometimes they recognize that more than one explanation fits their data. Then they try out their proposed explanations by making predictions that they test. At first I had you draw a picture of how you thought my tube worked and had you each explain your picture. You got to hear each other's view on how the system worked. Doug, did you change your ideas at all based on what you heard from Karen?"

"Well, yes. I first thought that ropes A and C were the two ends of the same rope and B and D were two ends of another rope. Karen had A and B as ends of the same rope and C and D as ends of another rope, and her explanation seemed to fit better than mine."

"Right. Communication about observations and interpretations is very important among scientists because different scientists may interpret data in different ways. Hearing someone else's views can help a scientist revise his or her interpretation. In essence that was what you were doing when you shared your diagrams. Karen, when your model didn't work, what did you do?"

"All I did was adjust the length of one rope, and then it worked fine."

"So as a result of your formal testing of the predictions from your model, you revised your explanation of the system. Your understanding improved. In scientific terms, you revised your model to make it more consistent with your further observations. In science, the validity of any explanation is determined by its coherence with observations in the natural world and by its ability to predict further observations."

"But we still have different models," Karen observes. "How do we know which one is right?"

Doug says: "You told us that, didn't you, Barbara. There can be two possible explanations for the same observation."

"So it's possible for scientists to disagree sometimes," says Karen. "But does that mean that we don't understand evolution because scientists disagree about how evolution takes place?"

"Not at all," Barbara answers, "you both created different models of my tube, but both of your models are fairly accurate. And don't forget there were constraints on the possible models you could create that would be consistent with the data. Just any explanation would not be acceptable.

In evolution, there are some things we know could not have happened, just as we are confident that some things have happened." "And if different scientists can have different explanations, like Karen and I did, then I guess science also has to involve judgment to some extent," Doug says. "But I thought scientists were supposed to be totally objective," says Karen. "Good science always attempts to be objective, but it also relies on the individual insights of scientists.

And the questions they choose to ask as well as the methods they choose to use, not to mention the interpretations they may have, can be colored by their individual interests and backgrounds. But scientific explanations are reviewed by other scientists and must be consistent with the natural world and future experiments, so there are checks on subjectivity.

What we read in science books is a combination of observations and inferred explanations of those observations that can change with new research." "Still, I'm wondering," says Karen, "how can we find out which model is right?" "Let's just open up Barbara's tube," says Doug.

"We could do that," Barbara says. "But let's assume in this analogy that opening the tube is not possible. Sometimes scientists figure out how to open up the natural world and look inside, but sometimes they can't.

And not opening up the tube is a good metaphor for how science often works. Science involves coming up with explanations that are based on evidence. With time, additional evidence might require changing the explanations, so that at any time what we have is the best explanation possible for how things work. In the future, with additional data, we may change our original explanation— just like you did, Karen.

"Remember when we were talking this morning about evolution being fact or theory? That conversation is very relevant to what we have been doing with the tubes. As scientists started to notice patterns in nature, they began to speculate about some explanations for these patterns.

These explanations are analogous to your initial ideas about how my tube worked. In the terms of science, these initial ideas are called hypotheses. You noticed some patterns in how the ropes were related to each other, and you used these patterns to develop a model to explain the patterns. The model you created is analogous to the beginning of a scientific theory. Except in science, theories are only formalized after many years of testing the predictions that come from the model.

"Because of our human limitations in collecting complete data, theories necessarily contain some judgments about what is important. Judgments aren't a weakness of scientific theory. They are a basic part of how science works."

"I always thought of science as a bunch of absolute facts," says Doug. "I never thought about how knowledge is developed by scientists."

"Creativity and insight are what help make science such a powerful way of understanding the natural world. "There's another important thing that I try to teach my students with this activity," Barbara continues. "It's important for them to be able to distinguish questions that can be answered by science from those that cannot be answered by science. Here's a list of questions that I use to get them talking. I ask them if a question can be answered by science, cannot be answered by science, or has some parts that belong to science and others that do not. Then I ask the group to select a couple of questions and discuss how they would go about answering them."

Barbara hands Doug and Karen the following list of questions:

Do ghosts haunt old houses at night?

How old is the earth?

Should I follow the advice of my daily horoscope?

Do species change over long periods of time?

Should I exercise regularly?

"Of course, you can make up other questions if something is happening in the news or if it's related to an earlier lesson. And sometimes I include moral or religious questions to make it clear that they lie outside science."

"I can see that these would get students thinking," says Karen. "I guess understanding the nature of science really is relevant to real life."

"That's what this exercise is about."

2

THE NATURE OF SCIENCE

Science is a particular way of knowing about the world. In science, explanations are restricted to those that can be inferred from confirmable data—the results obtained through observations and experiments that can be substantiated by other scientists. Anything that can be observed or measured is amenable to scientific investigation. Explanations that cannot be based on empirical evidence are not a part of science.

The history of life on earth is a fascinating subject that can be studied through observations made today, and these observations have led to compelling accounts of how organisms have changed over time. The best available evidence suggests that life on earth began more than three and a half billion years ago. For more than two billion years after that, life was housed in the bodies of many kinds of tiny, singlecelled organisms, some of which produced the oxygen that now makes up more than a fifth of the earth's atmosphere. Less than a billion years ago, much more complex organisms appeared. By about half a billion years ago, evolution had resulted in a wide variety of multicellular animals and plants living in the sea that are the clear ancestors of many of the major types of organisms that continue to live to this day. Somewhat more than 400 million years ago, some marine plants and animals began one of the greatest of all innovations in evolution— they invaded dry land. For our own phylum, the Chordata, this move away from the nurturing sea led to the appearance of amphibians, reptiles, birds, and mammals— the latter including, of course, our own species, *Homo sapiens*.

This chapter looks at how science works in the context of our overall understanding of how biological evolution occurred. It begins, however, by discussing another scientific development that challenged long-held understandings and beliefs: the discovery of heliocentricism.

HELIOCENTRICISM AND THE NATURE OF SCIENCE

Surely one of the first major natural phenomena to be understood was the cause of night and day. Some of the earliest surviving human records left on clay tablets relate to the movements of the sun and other celestial bodies. The obvious cause of day and night is the rising and setting of the sun. This is an observation that can be made today by anyone and, seemingly, requires no further explanation. Archaeological evidence and early records make it clear that our ancestors realized that not only does the sun appear to rise and set, but so do the moon and stars. The movements of the moon and stars, however, are not precisely synchronized with those of the sun. The moon is slower by about one hour per day.

The stars remain almost the same on successive nights, but slowly it becomes obvious that they, too, are slowed in their movements compared to the sun. Thus, the stars of summer are different from those visible in the winter. In fact, it takes a full year for the stars to return to their previous position, an interval of time that defines our year.

The ancient observers realized that not all stars move in unison. Although most move in majestic unity, a few others are "wanderers"—appearing now with one group of stars and a week later somewhere else. The majority were called "fixed stars," the wanderers were called "planets." During the late Middle Ages, and especially in the Renaissance, beautiful brass models known as orreries were made to show the relative positions and movements of the sun, planets, and moon as they circled the earth. As the center of the universe, the earth was a sphere in the center of the orrery. The other celestial bodies were positioned on rings of metal, each moving by clockwork at its own rate. The fixed stars required a simple solution—they could be considered stuck in an outermost shell, also moved by clockwork.

The problem with orreries—and with the theories of the cosmos then prevailing— was that they had to become successively more complex as more became known. Careful observations of the movements of the stars and planets greatly complicated the hypotheses used to account for those movements. This growing complexity stimulated some of the leading astronomers of the 16th and 17th centuries, including Copernicus, Kepler, and Galileo, to make even more precise observations of the movements of the heavenly bodies.

Astronomers used these measurements to demonstrate that the ageold human explanations of the heavens were incomplete. In the process, they replaced a complex and confusing explanation with a simple one: the sun, rather than the earth, is at the center of a "solar system," and the earth revolves around it. That simple step—a bold departure from past thinking due mainly to the insights of Copernicus (1473 to 1543)—dramatically changed the picture of the then known universe. This concept of heliocentricism initially ran counter to the positions of religious authorities.

The view of Christianity over most of its history, based on a literal inter-pretation of the Bible, was that the earth is the center of the universe around which the celestial bodies revolve. Copernicus dedicated his book describing the theory of heliocentricism, *De revolutionibus orbium coelestium*, to Pope Paul III and promptly died. That saved him the troubles that were to beset Galileo (1564 to 1642), whose astronomical observations confirmed the views of Copernicus.

Galileo was told to abandon his beliefs, and he later was tried by the Inquisition and sentenced to the equivalent of house arrest. The Church held that his views were dangerous to faith. Continued study and ever more careful measurements of the movements of the planets and sun continued to support the heliocentric hypothesis. Then, in the latter half of the 17th century, Isaac Newton (1642 to 1727) showed that the force of gravity—as measured on the earth—could account for the movements of the planets given the laws of motion that Newton derived. As a result of the steady accumulation of evidence, the theological interpretation of celestial movements gave way to the naturalistic explanation, and it is now accepted that night and day are the consequences of the rotation of the earth on its axis. Today, we can see for ourselves the rotation of the earth from satellites orbiting the planet.

Like biological evolution, the theory of heliocentricism brought order and new understanding to an otherwise chaotic and confusing aspect of nature. It also had great practical applications, in that the exploration of the world by European seafarers used the more accurate understanding of celestial mechanics to assist in navigation. Looking at the night sky remains a powerful experience. But that experience is now informed not only by the beauty and majesty of the heavens, but by a deeper understanding of nature and by an appreciation of the power of the human intellect. This triumph of the human mind says a great deal about the nature of science. First, science is not the same as common sense. Common sense indicates that the sun does rise and set. Nevertheless, there can be other explanations of that phenomenon, and one of them, the rotation of the earth on its axis, is responsible for day and night. A concept based on observation proved to need extensive modification as new observations accumulated.

Second, the statements of science should never be accepted as "final truth." Instead, over time they generally form a sequence of increasingly more accurate statements. Nevertheless, in the case of heliocentricism as in evolution, the data are so convincing that the accuracy of the theory is no longer questioned in science. Third, scientific progress depends on individuals, but the contributions of one individual could be made by others. If Copernicus had kept his ideas to himself, the discovery of heliocentricism would have been postponed, but it would not have been blocked, since other astronomers eventually would have come to the same conclusion. Similarly, had Darwin and Wallace not published their hypotheses, the concept of biological evolution would nevertheless have emerged as the accepted explanation for the history of life on earth. The same cannot be said in other areas of human endeavor; for example, had Shakespeare never published, we would most assuredly never have had his plays. The publications of scientists, unlike those of playwrights, are a means to an end—they are not the end itself.

SCIENCE REQUIRES CAREFUL DESCRIPTION

What are the scientific methods that have led to our current understanding of the history of life over vast eons of time? They begin with careful descriptions of the material being studied.

The material for the study of biological evolution is life itself. One basic aspect of life is that individuals can be grouped as similar kinds, or species. Another important observation is that many species seem to be closely related to each other. The scientific classification of species and their arrangement into groups began with the publication in 1758 of *Systema Naturae*, or system of nature, by the Swedish naturalist Carolus Linnaeus (1707 to 1778). For example, Linnaeus knew seven dog-like species, and he gave each a double name. Subsequently many more species were discovered and some of the names were changed—and continue to be changed as more information is obtained. The domestic dog is *Canis familiaris*; the coyote of North America is *Canis latrans*; the Australian dingo is *Canis dingo*; and the wolf of the northern hemisphere is *Canis lupus*. Thus *Canis* is the name of the genus of dog-like animals, and the distinctive second name is the species name.

Generations of scientists have discovered new species, described them, and arranged them into the system first suggested by Linnaeus. Whereas Linnaeus recognized about 9,000 species, systematists now have recognized about 1.5 million. The task of categorizing and describing species is still far from complete. Most species of smaller invertebrates, and many bacteria and other microscopic organisms, remain to be discovered. The plant kingdom is also incompletely known. Though the flowering plants of many areas, such as Europe and North America, are fairly well described, many other regions have not been nearly as well explored by botanists.

Recent investigations in the exceptionally diverse rainforests of South America have caused biologists to raise their estimates of the number of undescribed species. For example, a very high proportion of insects collected from the forest canopy are "new" species to science. It is now believed that the number of different species of plants and animals in the world may be ten million, or even more.

The scientific methods used in classifying organisms have been greatly improved over time. The process begins with the intensive field work in which the animals, plants, and microorganisms are collected and carefully examined. Most will be known to a specialist, but there might be some unusual examples. However, none is likely to be a complete stranger, since the specialist will probably recognize that any puzzling specimen is similar to some familiar species. Next the specialist must check all that has been published on the group of organisms that contains the similar species. If, after an exhaustive search, there is no record of a described species that corresponds to the one being examined, the specimen is probably a new species. The specialist will then prepare a careful description of the new species and publish it in a scientific journal. There is a permanent reward for being the describer of a new species: thereafter monographs that deal with the classification of the group to which the new species belongs will add the describer's name at the end of the scientific name. Thus, for example, "*Homo sapiens* Linnaeus" is our own proper identification, because Linnaeus was the first to give us our scientific name.

This example makes it clear that not all scientific data are derived as the result of experimentation. The conventional classification of species into seemingly natural groups involved the careful observation of a variety of different species, followed by the use of selected characteristics in an attempt to define groups of species thought to be related. But the groupings are not always obvious. For example, it might have seemed reasonable to classify bats with birds, since the most conspicuous characteristic of each is the ability to fly. But bats are mammals. Like all mammals, their bodies are covered with hair and their young are born alive (instead of hatching from eggs) and are nourished by milk from the mother's mammary glands.

Although most of the species we know today were described after the time of Linnaeus, we continue to use his basic system of hierarchical classification. For example, similar genera are united in families, similar families in orders, similar orders in classes, and similar classes in phyla. The dog-like species listed above (the genus *Canis),* plus a number of similar but more distant dog-like animals, are grouped as the family Canidae. This family plus the families of cats, bears, seals, and weasels form the order Carnivora. The carnivores and all other animals with hair are combined as the class Mammalia. Mammals are combined with the birds, reptiles, amphibians, fishes, plus some small marine animals in the phylum Chordata. Today, many systematists group organisms according to a system known as cladistics. By determining which traits of a species evolved earlier and which evolved later, this system seeks to classify organisms according to their evolutionary history.

SCIENCE AS EXPLANATION

In the quest for understanding, science involves a great deal of careful observation that eventually produces an elaborate written description of the natural world. This description is communicated to scientists in scientific journals or at scientific meetings, so that others can build on pre-existing work. In this way, the accuracy and sophis-tication of the description tends to increase with time, as subsequent generations of scientists correct and extend the observations of their predecessors. Because the total sum of scientific knowledge increases relentlessly, scientific progress is something that all scientists take for granted. But science is not just description.

Even as observations are being made, the human mind attempts to sort, or organize, the observations in a way that reveals some underlying order in the objects or phenomena being observed. This sorting process, which involves a great deal of trial and error, seems to be driven by a fundamental human urge to make sense of our world. The sorting process also suggests new observations that might otherwise not be made. For example, the suggestion that bats should be grouped with mammals led to an intensified examination of the similarities between bats and rodents—first at the anatomical level, and later with respect to the genes and protein molecules that form their cells. In this case, new evidence was obtained that confirmed the suggested relationship.

In other cases, the further observations inspired by a tentative grouping have caused the rejection of a new idea. The realization that species can be arranged in a hierarchy of groups of seemingly similar forms raised an obvious question: What accounts for the relatedness of different groups of organisms? The mechanism that was proposed by Darwin directly addressed this question. It suggested that all animals classified as belonging to the same group had a common ancestor species. That is, dogs, wolves, coyotes and all members of the genus *Canis* are descended from a common ancestor species that lived in the remote past. In a similar manner all species in a family, an order, a class, or a phylum share a common inheritance.

How could one possibly test such a hypothesis? In the decades before Darwin proposed his hypothesis, geologists realized that the sedimentary rocks of the earth's crust contain a running diary of earth's history. This record of past events comes about because the earth's crust is in a con-stant state of change. This observation might not be obvious in the lifetime of an individual, but it is dramatic over thousands of years. Relatively flat surfaces are uplifted to form mountains, and then the mountains slowly erode to form flatlands. Storms produce powerful waves that erode cliffs at the seashore. These phenomena have the common feature of moving solid materials, and the subsequent settling out of these materials makes possible the formation of a special form of rock that contains a record of the earth's past.

Consider the case of a river with a source in the mountains. As the water moves downstream, it erodes the slopes of the mountains. Tiny grains produced by the erosion, called silt, are relatively easy to move. When the river reaches the flatlands, a lake, or the ocean, the solid material being carried by the water is deposited—often reaching great thicknesses over long periods of time. Then the pressure of the sediments on top can cause the sediments beneath to harden into "sedimentary rocks." The river may carry things other than silt, sand, and rocks. Hard structures of organisms such as the bones and teeth of animals may be carried along as well. These, too, will be deposited with the silt, sand, and rocks. Under certain circumstances, these remains of organisms undergo a chemical change in which the original material is replaced by molecules that form stone. In this way, the organic remains of living things are fossilized (changed into stone), creating the evidence of ancient life studied by scientists.

Because of the order in which the sediments are deposited, the most recent layer of rocks normally will be on top and the oldest layer will be on the bottom (though sometimes sediments are flipped upside down by the geologic folding of rock layers). Also, the fossils in each layer usually will be of those organisms that lived at the time the layer was formed. Thus, the fossils in the lower layers will represent species that lived earlier than those found in the upper layers.

The relative position of fossils tells only which are older and which younger. One can estimate the difference in the ages of the two fossils by noting the thickness of the rock that separates them. If the difference is only one foot, one might guess the interval of time is less than if two fossils are separated by 50 feet of rock layers. Today, how-ever, far more accurate methods of dating fossils are available, as described on the next page. Because these methods are based on the known rates of radioactive decay, they provide valuable measures of absolute time. The scientific study of fossils is called paleontology, and the methods used for their identification and classification are similar to those used for living species. But in some respects the task of the paleontologist is far more difficult. Many species lack hard parts such as bones and shells, and such organisms almost always decay without becoming fossilized. This is the case for many groups of soft-bodied invertebrates— such as worms of many kinds, jellyfish, and protozoans. Even for such species as mammals, birds, reptiles, and amphibians, death is usually followed by the skeleton being dismembered and the bones scattered. For this reason, whereas isolated bones are often fossilized, it is exceptionally rare for an intact skeleton to be found.

Tiny fossils first reveal the existence of bacteria 3.5 to 3.8 billion years ago, and animals composed of more than a single cell are known from about 670 million years ago. But the organisms that lived between these two dates lacked hard parts and, hence, were rarely preserved as fossils. Then, about 570 million years ago, a dramatic change took place. At the beginning of the Cambrian period, animals evolved that had calcified shells and other types of body coverings that had a far better chance of becoming fossilized. These fossils demonstrate that Cambrian seas were populated with a variety of invertebrates. The earliest vertebrate fossils date from about 500 million years ago. Thereafter early amphibians and reptiles appeared. Birds and mammals appear in the fossil record only about 200 million years ago, while dinosaurs first appear about 225 million years ago and disappear suddenly about 160 million years later.

In the 1830s, when Darwin began his studies, the essential features of the fossil record were known (although absolute dates had not yet been determined). Many thousands of living species had been described, and it was clearly recognized that they could be organized into various groups—suggesting that they are somehow relatives. In addition, analysis of the fossil record revealed that the organisms on the ancient earth had undergone major changes over time—with whole groups of animals appearing, persisting for long periods of time, and then disappearing.

Dating the Earth

One of the greatest scientific triumphs of the last two centuries has been the discovery of the vast expanse of geologic time. Early methods of calculating the age of the earth relied on measures of the rate of sedimentation or the cooling of the earth from an initially molten state. The relative ages of rocks also were calculated early in the 1800s by noting what kinds of fossils the rocks contained. But the absolute age of the earth and the timing of many events in geologic history required the discovery late in the 19th century of a previously unknown phenomenon: radioactivity.

Some elements, such as uranium, undergo radioactive decay to produce other elements. By measuring the quantities of radioactive elements and the elements into which they decay in rocks, geologists can determine how much time has elapsed since the rock cooled from an initially molten state. For example, the oldest known rocks are found in Greenland and date from about 3.8 billion years ago. Scientists believe the earth's age to be about 4.6 bil-lion years because meteorites and rocks of the moon—both of which formed about the same time as the earth—date from this time. Radiometric dating also shows that the period of earth's history during which large fos-sils can be readily found in rocks began only about 570 million years ago.

Radiometric dating draws on information and insights from many areas of science. For example, it requires that the rate of radioactive decay is constant over time and is not influenced by such factors as temperature or pressure—conclusions supported by extensive research in physics. It also assumes that the rocks being analyzed have not been altered over time by the migration of atoms into or out of the rocks, which requires detailed information from both the geologic and chemical sciences.

Darwin was an unusually keen observer. But he was not content to catalogue facts and observations. Instead, the natural world to him was a gigantic, very challenging puzzle that demanded an explanation for its otherwise bewildering complexity. Why are different organisms so similar? Why has there been a succession of different kinds of species throughout geologic time? Certain observations seemed particularly important. For example:

1) In South America, the only continent where living armadillos were found, Darwin discovered fossil evidence for the prior existence of ancient species that had many of the unique features of living armadillos, yet were clearly different. Such fossils were found nowhere else in the

world. Why were both living and ancient armadillo-like species confined to the same geographical region?

2) On the Galapagos Islands, 600 miles off the coast of Ecuador, Darwin observed many distinct living species of birds and reptiles that closely resembled each other—yet were different on each tiny island. Why, for example, should the beak size of the mockingbirds on one island be different from that of a closely related mockingbird on an island only 30 miles away? And why were the various types of animals on these islands related, but distinct from, the animals in Ecuador, whereas those on the otherwise very similar islands off the coast of Africa were related to the animals in Africa instead?

Darwin could not see how these observations could be explained by the prevailing view of his time: that each species had been independently created, with the species that were best suited to each location on the earth being created at each particular site. It looked instead as though species could evolve from one into another over time, with each being confined to the particular geographical region where its ancestors happened to be—particularly if isolated by major barriers to migration, such as vast expanses of ocean.

But how could one species turn into another over the course of time? In constructing his hypothesis of how this occurred, Darwin was struck by several other observations that he and others before him had made.

1) People who bred domesticated animals and plants for commercial or recreational use had found and exploited a great deal of variation among the progeny of their crosses. Pigeon breeders, for example, had observed wide differences in colors, beaks, necks, feet, and tails of the offspring from a single mating pair. They rou-tinely enhanced their stocks for desired traits—for example, selectively breeding those animals that shared a particular type of beak. Through such artificial selection, pigeon fanciers had been able to create many different-looking pigeons, known as breeds. A similar type of artificial

selection for mating pairs of dogs had likewise created the whole variety of shapes and sizes of these common pets—ranging from a Great Dane to a dachshund.

2) Animals living in the wild can face a tremendous struggle for survival. For some birds, for example, fewer than one in 100 animals born in one year will survive over a harsh winter into a second year. Those with characteristics best suited for a particular environment—for example, those individual birds who are best able to find scarce food in the winter while avoiding becoming food for a larger animal—tend to have better chances of surviving. Darwin called this process natural selection to distinguish it from the artificial selection used by dog and pigeon breeders to determine which animals to mate to produce offspring.

At least 20 years elapsed between the time that Darwin conceived of descent with modification and 1859, the year that he revealed his ideas to the world in *On the Origin of Species*. Throughout these 20 years, Darwin did what scientists today do: he tested his ideas of how things work with new observations and experiments. In part, he did this by thinking up every possible objection he could to his own hypothesis.

For each such argument, Darwin tried to find an observation made by others, make an observation, or do an experiment of his own that might imply that his ideas were in fact not valid. When he could successfully counter such objections, he strengthened his theory.

For example, Darwin's ideas readily explained why distant oceanic islands were generally devoid of terrestrial mammals, except for flying bats. But how could the land snails, so common on such islands, have traversed the hundreds of miles of open ocean that separate the islands from the mainland where the snails first evolved?

By floating snails on salt-water for prolonged periods, Darwin convinced himself that, on rare occasions in the past, snails might in fact have "floated in chunks of drifted timber across moderately wide arms of the sea." This example shows how a hypothesis can drive a scientist to do experiments that would otherwise not be done.

Prior to Darwin, the existence of land snails and bats, but not typical terrestrial mammals, on the oceanic islands was simply noted and catalogued as a fact. It is unlikely that anyone would have thought to test the snails for their ability to survive for prolonged periods in salt water. Even if they had, such an experiment would have had little meaning or impact.

By publishing his ideas, Darwin subjected his hypothesis to the tests of others. This process of public scrutiny is an essential part of science.

It works to eliminate individual bias and subjectivity, because others must also be able to determine whether a proposed explanation is consistent with the available evidence. It also leads to further observations or to experiments designed to test hypotheses, which has the effect of advancing science.

Many of the hypotheses advanced by scientists turn out to be incorrect when tested by further observations or experiments. But skillful scientists like Darwin tend to have good ideas that end up increasing the amount of knowledge in the world. For this reason, the ideas of scientists have been—over the long run—central to much of human progress.

Continental Drift and Plate Tectonics: A Scientific Revolution of the Past 50 Years

The theory of plate tectonics demonstrates that revolutions in science are not just a thing of the past, thus suggesting that more revolutions can be expected in the future. World maps have long indicated a curious "jigsaw puzzle fit" of the continents. This is especially apparent between the facing coastlines of South America and Africa. Alfred Wegener (1880 to 1930), a German meteorologist who was dissatisfied with explanations that relied on expanding and contracting crust to account for mountain building and the formation of the ocean floor, pursued other lines of reasoning. Wegener suggested that all of earth's conti-nents used to be assembled in a single ancient super-continent he called Pangea. He hypothesized that Pangea began to break up approximately 200 million years ago, with South America and Africa slowly drifting apart to their present positions, leaving the southern Atlantic Ocean between them. This was an astonishing hypothesis: could huge continents really move?

Wegener cited both geological and biological evidence in support of his explanation. Similar plant and animal fossils are found in rock layers more than 200 million years old in those regions where he claimed that different continents were once aligned. Wegener attributed this to the migration of plants and animals freely throughout these broad regions. If 200 million years ago Africa and South America had been separated by the Atlantic Ocean as they are today, their climates, environments, and life forms should have been very different from each other—but they were not. Despite Wegener's use of evidence and logic to develop his explanations, other scientists found it difficult to imagine how solid, brittle continents could plow through the equally solid and brittle rock material of the ocean floor. Wegener did not have an explanation for how the continents moved. Since there was no plausible mechanism for continental drift, the idea did not take hold. The hypothesis of continental drift was equivalent to the hypothesis of evolution in the decades before Darwin, when evolution lacked the idea of variation followed by natural selection as an explanatory mechanism.

The argument essentially lay dormant until improved technologies allowed scientists to gather previously unobtainable data. From the mid 1950s through the early 1970s, new evidence for a mechanism to explain continental drift became available that the scientific community could accept. Sonar mapping of the ocean floor revealed the presence of a winding, continuous ridge system around the globe. These ridges were places where molten material was welling up from the earth's interior and pushing apart the plates that form the earth's surface.

SCIENCE ASCUMULATIVE KNOWLEDGE

At the time of Darwin, there were many unsolved puzzles, including missing links in the fossil record between major groups of animals. Guided by the central idea of evolution, thousands of scientists have spent their lives searching for evidence that either supports or conflicts with the idea. For example, since Darwin's time, paleontolo-gists have discovered many ancient organisms that connect major groups—such as *Archaeopteryx* between ancient reptiles and birds, and *Ichthyostega* between ancient fish and amphibians. By now, so much evidence has been found that supports the fundamental idea of biological evolution that its occurrence is no longer questioned in science.

Even more striking has been the information obtained during the 20th century from studies on the molecular basis of life. The theory of evolution implies that each organism should contain detailed molecular evidence of its relative place in the hierarchy of living things. This evidence can be found in the DNA sequences of living organisms.

Before a cell can divide to produce two daughter cells, it must make a new copy of its DNA. In copying its DNA nucleotides, however, cells inevitably make a small number of mistakes. For this reason, a few nucleotides are changed through random error each time that a cell divides. (For example, an A in the DNA sequence of a gene in a chromosome may be replaced with a G in the new copy made as the cell divides.) Therefore, the larger number of cell divisions that have elapsed between the time that two organisms diverged from their common ancestor, the more differences there will be in their DNA sequences due to chance errors.

This molecular divergence allows researchers to track evolutionary events by sequencing the DNA of different organisms. For example, the lineage that led to humans and to chimpanzees diverged about 5 million years ago—whereas one needs to look back in time about 80 million years to find the last common ancestor shared by mice and humans. As a result, there is a much smaller difference between human and chimpanzee DNA than between human (or chimpanzee) and mouse DNA. In fact, scientists today routinely use the differences they can measure between the DNA sequences of organisms as "molecular clocks" to decipher the relationships between living things.

The same comparisons among organisms can be made using the proteins encoded by DNA.

For example, every living cell uses a protein called cytochrome c in its energy metabolism. The cytochrome c proteins from humans and chimpanzees are identical. In a relatively short time, these new observations, measurements, and interpretations provoked a complete shift in the thinking of the scientific community. Geologists now accept the idea that the surface of the earth is broken up into about a dozen large pieces, as well as a number of smaller ones, called tectonic plates.

On a time scale of millions of years, these plates shift about on the planet's surface, changing the relative positions of the continents.

The plate tectonic model provides. But there is only an 86 percent overlap in the molecules between humans and rattlesnakes, and only a 58 percent overlap between us and brewer's yeast. This is explained by the evolutionary proposition that we shared a common ancestor with chimps relatively recently, whereas the com-mon ancestor that we, as vertebrates, shared with rattlesnakes is much more ancient. Still farther in the past, we and yeast shared a common ancestor—and the molecular data reflect this pattern.

In the past few decades, new methods have been developed that are allowing us to explanations that are widely accepted for the evolution of crustal features such as folded mountain chains, zones of active volcanoes and earthquakes, and deep ocean floor trenches. Direct measurements using the satellite-based global positioning system (GPS) to measure absolute longitude and latitude verify that the plates collide, move apart, and slide past one another in differ-ent areas along their adjacent boundaries at speeds comparable to the growth rate of a human fingernail.

· Obtain the exact sequence of all of the DNA nucleotides in chromosomes. The Human Genome Project, for example, will produce when completed the entire sequence of the 3 billion nucleotides that make up our genetic inheritance. The complete sequence of the yeast genome (12 million nucleotides) is already known, as are the genomes for numerous species of bacteria (from 0.5 to 5 million nucleotides each, depending on the species). Similar sequencing efforts will soon yield the complete sequences for hundreds of bacteria and other organisms with small genomes.

These molecular studies are powerful evidence for evolution. The exact order of the genes on our chromosomes can be used to predict the order on monkey or even mouse chromosomes, since long stretches of the chromosomes of mammalian species are so similar. Even the parts of our DNA that do not code for proteins and at this point have no known function are similar to the comparable parts of DNA in related organisms.

The confirmation of Darwin's ideas about "descent with modification" by this recent molecular evidence has been one of the most exciting developments in biology in this century. In fact, as the chromosomes of more and more organisms are sequenced over the next few decades, these data will be used to reconstruct much of the missing history of life on earth—thereby compensating for many of the gaps that still remain in the fossil record.

CONCLUSION

One goal of science is to understand nature. "Understanding" in science means relating one natural phenomenon to another and recognizing the causes and effects of phenomena. Thus, scientists develop explanations for the changing of the seasons, the movements of heavenly bodies, the structure of matter, the shaping of mountains and valleys, the changes in the positions of continents over time, and the diversity of living things.

The statements of science must invoke only natural things and processes. The statements of science are those that emerge from the application of human intelligence to data obtained from observation and experiment. These fundamental characteristics of science have demonstrated remarkable power in allowing us to describe the natural world accurately and to identify the underlying causes of natural phenomena. This understanding has great practical value, in part because it allows us to better predict future events that rely on natural processes.

Progress in science consists of the development of better explanations for the causes of natural phenomena. Scientists can never be sure that a given explanation is complete and final. Yet many scientific explanations have been so thoroughly tested and confirmed that they are held with great confidence.

The theory of evolution is one of these explanations. An enormous amount of scientific investigation has converted what was initially a hypothesis into a theory that is no longer questioned in science. At the same time, evolution remains an extremely active field of research, with an abundance of new discoveries that are continually increasing our understanding of exactly how the evolution of living organisms actually occurred.

3

NATIONAL SCIENCE EDUCATION STANDARDS

Over the last six years, several major documents have been released that describe what students from kindergarten through twelfth grade should know and be able to do as a result of their instruction in the sciences. These include the *National Science Education Standards* released by the National Research Council in 1996,[1] the Benchmarks for Science Literacy released by the American Association for the Advancement of Science in 1993,[2] and *The* Content Core: A Guide for Curriculum Designers released by the Scope, Sequence, Coordination project of the National Science Teachers Association in 1992.[3]

These documents agree that all students should leave biology class with an understanding of the basic concepts of biological evolution and of the limits, possibilities, and dynamics of science as a way of knowing. *Benchmarks for Science Literacy*, for example, states that "the educational goal should be for all children to understand the concept of evolution by natural selection, the evidence and arguments that support it, and its importance in history." For biology educators, these documents offer significant support for the inclusion of evolution in school science programs.

STRUCTURE AND OVERVIEW OF THE NATIONAL SCIENCE EDUCATION STANDARDS

This chapter focuses on the treatment of evolution in the *National Science Education Standards*. The *Standards* are divided into six broad sections. The first set of standards, the *science teaching standards,* describes what teachers of science at all grade levels should know and be able to do. The professional development standards describe the experiences necessary for teachers to gain the knowledge, understanding, and ability to implement the *Standards*. The *assessment standards* provide criteria against which to judge whether assessments are contributing fully to the goals outlined in the *Standards*. The *science content standards* outline what students should know, under-stand, and be able to do in the natural sciences. The *science education program standards* discuss the planning and actions needed to translate the *Standards* into programs that reflect local contexts and policies. And the *science education system standards* consist of criteria for judging the performance of the overall science education system.

The *Standards* rest on the premise that science is an active process. Learning science is something that students do, not something that is done to them. "Hands-on" activities, although essential, are not enough. Students must have "minds-on" experiences as well.

The *Standards* make inquiry a central part of science learning. When engaging in inquiry, stu-dents describe objects and events, ask questions, construct explanations, test those explanations against current scientific knowledge, and communicate their ideas to others. They identify their assumptions, use critical and logical thinking, and consider alternative explanations. In this way, students actively develop their understanding of science by combining scientific knowledge with reasoning and thinking skills.

The importance of inquiry does not imply that all teachers should pursue a single approach to teaching science. Just as inquiry has many different facets, so too do teachers need to use many different strategies to develop the understandings and abilities described in the *Standards*.

Nor should the *Standards* be seen as requiring a specific curriculum. A curriculum is the way content is organized and presented in the classroom. The content embodied in the *Standards* can be organized and presented with different emphases and perspectives in many different curricula.

EVOLUTION AND THE NATURE OF SCIENCE IN THE NATIONAL SCIENCE EDUCATION STANDARDS

Evolution and the nature of science are major topics in the content standards. The first mention of evolution is in the initial content standard, entitled "Unifying Concepts and Processes." This standard points out that conceptual and procedural schemes unify science disciplines and provide students with powerful ideas to help them understand the natural world. It is the only standard that extends across all grades, because the understanding and abilities associated with this standard need to be developed over an entire education.

The standard is as follows:

As a result of activities in grades K–12, all students should develop understanding and abilities aligned with the following concepts and processes:

- *Systems, order, and organization*
- *Evidence, models, and explanation*
- *Constancy, change, and measurement*
- *Evolution and equilibrium*
- *Form and function*

The guidance offered for the standard is to establish a broad context for thinking about evolution: Evolution is a series of changes, some gradual and some sporadic, that accounts for the present form and function of objects, organisms, and natural and designed systems. The general idea of evolution is that the present arises from materials and forms of the past. Although evolution is most commonly associated with the biological theory explaining the process of descent with modification of organisms from common ancestors, evolution also describes changes in the universe.

With this unifying standard as a basis, the remaining content standards are organized by age group and discipline.

Grades K–4

The life science standard for grades K–4 is organized into the categories of characteristics of organisms, life cycles of organisms, and organisms and their environments. Evolution is not explicitly mentioned in these standards, but the text explains the basic things in life science that elementary school children ought to be able to understand and do: During the elementary grades, children build understanding of biological concepts through direct experience with living things, their life cycles, and their habitats.

These experiences emerge from the sense of wonder and natural interests of children who ask questions such as: "How do plants get food? How many different animals are there? Why do some animals eat other animals? What is the largest plant? Where did the dinosaurs go?" An understanding of the characteristics of organisms, life cycles of organisms, and of the complex interactions among all components of the natural environment begins with questions such as these and an understanding of how individual organisms maintain and continue life.

The intention of the K–4 standard is to develop the knowledge base that will be needed when the fundamental concepts of evolution are introduced in the middle and high school years.

Grades 5–8

For grades 5–8, the life science standard is the following:

As a result of their activities in grades 5–8, all students should develop understanding of:

- *Structure and function in living systems*
- *Reproduction and heredity*
- *Regulation and behavior*
- *Populations and ecosystems*
- *Diversity and adaptations of organisms*

The guidance for this standard defines regulation and behavior as follows:

An organism's behavior evolves through adaptation to its environment. How a species moves, obtains food, reproduces, and responds to danger are based in the species' evolutionary history. The text discusses diversity and adaptations as follows:

Diversity and Adaptations of Organisms

Millions of species of animals, plants, and microorganisms are alive today. Although different species might look dissimilar, the unity among organisms becomes apparent from an analysis of internal structures, the similarity of their chemical processes, and the evidence of common ancestry. Biological evolution accounts for the diversity of species developed through gradual processes over many generations.

Species acquire many of their unique characteristics through biological adaptation, which involves the selection of naturally occurring variations in populations. Biological adaptations include changes in structures, behaviors, or physiology that enhance survival and reproductive success in a particular environment.

Extinction of a species occurs when the environment changes and the adaptive characteristics of a species are insufficient to allow its survival. Fossils indicate that many organisms that lived long ago are extinct. Extinction of species is common; most of the species that have lived on the earth no longer exist. The text accompanying the standard also discusses some of the difficulties encountered in teaching about adaptation:

Understanding adaptation can be particularly troublesome at this level. Many students think adaptation means that individuals change in major ways in response to environmental changes (that is, if the environment changes, individual organisms deliberately adapt).

In fact, as described in Chapter 2 of this book, adaptation occurs through natural selection, a topic described under the life science standards for grades 9–12.

The content standards also treat evolution in grades 5–8 in the section on earth's history. The standard reads as follows:

As a result of their activities in grades 5–8, all students should develop an understanding of:

- *Structure of the earth system*
- *Earth's history*
- *Earth in the solar system*

The text discusses the importance of teaching students about earth systems and their interactions. A major goal of science in the middle grades is for students to develop an understanding of earth and the solar system as a set of closely coupled systems. The idea of systems provides a framework in which students can investigate the four major interacting *components of the earth system—geosphere* (crust, mantle, and core), hydrosphere (water), atmosphere (air), and the biosphere (the realm of all living things). In this holistic approach to studying the planet, physical, chemical, and biological processes act within and among the four components on a wide range of time scales to change continuously earth's crust, oceans, atmosphere, and living organisms. Their study of earth's history provides students with some evidence about co-evolution of *the planet's main features—the distribution of land* and sea, features of the crust, the composition of the atmosphere, global climate, and populations of living organisms in the biosphere.

The material offering guidance for the standard explicitly ties the earth's history to the history of life:

Earth's History

The earth processes we see today, including ero-sion, movement of lithospheric plates, and changes in atmospheric composition, are similar to those that occurred in the past. Earth's history is also influenced by occasional catastrophes, such as the impact of an asteroid or comet.

Fossils provide important evidence of how life and environmental conditions have changed. The standards for grades 5–8 cover the nature of science in the section on the history and nature of science:

As a result of activities in grades 5–8, all students should develop an understanding of:

- *Science as a human endeavor*
- *Nature of science*
- *History of science*

The guidance accompanying this standard offers the following discussion of these issues:

Nature of Science

Scientists formulate and test their explanations of nature using observation, experiments, and theoretical and mathematical models. Although all scientific ideas are tentative and subject to change and improvement in principle, for most major ideas in science, there is much experimental and observational confirmation. Those ideas are not likely to change greatly in the future. Scientists do and have changed their ideas about nature when they encounter new experimental evidence that does not match their existing explanations.

In areas where active research is being pursued and in which there is not a great deal of experimental or observational evidence and understanding, it is normal for scientists to differ with one another about the interpretation of the evidence or theory being considered. Different scientists might publish conflicting experimental results or might draw different conclusions from the same data. Ideally, scien-tists acknowledge such conflict and work towards finding evidence that will resolve their disagreement. It is part of scientific inquiry to evaluate the results of scientific investigations, experiments, observations, theoretical models, and the explanations proposed by other scientists.

Evaluation includes reviewing the experimental procedures, examining the evidence, identifying faulty reasoning, pointing out statements that go beyond the evidence, and suggesting alternative explanations for the same observations. Although scientists may disagree about explanations of phenomena, about interpretations of data, or about the value of rival theories, they do agree that questioning, response to criticism, and open communication are integral to the process of science. As scientific knowledge evolves, major disagreements are eventually resolved through such interactions between scientists.

History of Science

Many individuals have contributed to the traditions of science. Studying some of these individuals provides further understanding of scientific inquiry, science as a human endeavor, the nature of science, and the relationships between science and society. In historical perspective, science has been practiced by different individuals in different cultures.

In looking at the history of many peoples, one finds that scientists and engineers of high achievement are considered to be among the most valued contributors to their culture.

Tracing the history of science can show how difficult it was for scientific innovators to break through the accepted ideas of their time to reach the conclusions that we currently take for granted.

Grades 9–12

The life science standard for grades 9–12 directly addresses biological evolution. The standard reads as follows:

As a result of their activities in grades 9–12, all students should develop an understanding of:

- *The cell*
- *Molecular basis of heredity*
- *Biological evolution*
- *Interdependence of organisms*
- *Matter, energy, and organization in living systems*
- *Behavior of organisms*

The guidance for the life science standard describes the major themes of evolutionary theory:

Biological Evolution

Species evolve over time. Evolution is the consequence of the interactions of (1) the potential for a species to increase its numbers, (2) the genetic variability of offspring due to mutation and recombination of genes, (3) a finite supply of the resources required for life, and (4) the ensuing selection by the environment of those offspring better able to survive and leave offspring.

The great diversity of organisms is the result of more than 3.5 billion years of evolution that has filled every available niche with life forms. Natural selection and its evolutionary consequences provide a scientific explanation for the fossil record of ancient life forms, as well as for the striking molecular similarities observed among the diverse species of living organisms.

The millions of different species of plants, animals, and microorganisms that live on earth today are related by descent from common ancestors. Biological classifications are based on how organisms are related. Organisms are classified into a hierarchy of groups and subgroups based on similarities which reflect their evolutionary relationships. Species is the most fundamental unit of classification.

The text following the standard describes some of the difficulties that students can have in comprehending the basic concepts of evolution. Students have difficulty with the fundamental concepts of evolution.

For example, students often do not understand natural selection because they fail to make a conceptual connection between the occurrence of new variations in a population and the potential effect of those variations on the long-term survival of the species. One misconception that teachers may encounter involves students attributing new variations to an organism's need, environmental con-ditions, or use.

With some help, students can under-stand that, in general, mutations occur randomly and are selected because they help some organisms survive and produce more offspring. Other misconceptions center on a lack of understanding of how a population changes as a result of differential reproduction (some individuals producing more offspring), as opposed to all individuals in a population changing. Many misconceptions about the process of natural selection can be changed through instruction. Finally, evolution is discussed again in the guid-ance following the earth and space science standard:

As a result of their activities in grades 9–12, all students should develop an understanding of:

- *Energy in the earth system*
- *Geochemical cycles*
- *Origin and evolution of the earth system*
- *Origin and evolution of the universe*

The discussions of the origin and evolution of the earth system and the universe relate evolution to universal physical processes:

The Origin and Evolution of the Earth System The sun, the earth, and the rest of the solar system formed from a nebular cloud of dust and gas 4.5 billion years ago. The early earth was very different from the planet we live on today.

Geologic time can be estimated by observing rock sequences and using fossils to correlate the sequences at various locations. Current methods include using the known decay rates of radioactive isotopes present in rocks to measure the time since the rock was formed.

Interactions among the solid earth, the oceans, the atmosphere, and organisms have resulted in the ongoing evolution of the earth system. We can observe some changes such as earthquakes and volcanic eruptions on a human time scale, but many processes such as mountain building and plate movements take place over hundreds of millions of years.

Evidence for one-celled forms of life—the bacteria— extends back more than 3.5 billion years. The evolution of life caused dramatic changes in the composition of the earth's atmosphere, which did not originally contain oxygen.

The Origin and Evolution of the Universe

The origin of the universe remains one of the greatest questions in science. The "big bang" theory places the origin between 10 and 20 billion years ago, when the universe began in a hot dense state; according to this theory, the universe has been expanding ever since.

Early in the history of the universe, matter, primarily the light atoms hydrogen and helium, clumped together by gravitational attraction to form countless trillions of stars. Billions of galaxies, each of which is a gravitationally bound cluster of billions of stars, now form most of the visible mass in the universe.

Stars produce energy from nuclear reactions, primarily the fusion of hydrogen to form helium. These and other processes in stars have led to the formation of all the other elements.

The standard for the history and nature of science elaborates on the knowledge established in previous years:

As a result of activities in grades 9–12, all students should develop an understanding of:

- *Science as a human endeavor*
- *Nature of scientific knowledge*
- *Historical perspectives*

The discussion of this standard relates the nature of science explicitly to many of the problems that arise in the teaching of evolution.

Nature of Scientific Knowledge

Science distinguishes itself from other ways of knowing and from other bodies of knowledge through the use of empirical standards, logical arguments, and skepticism, as scientists strive for the best possible explanations about the natural world. Scientific explanations must meet certain criteria. First and foremost, they must be consistent with experimental and observational evidence about nature, and must make accurate predictions, when appropriate, about systems being studied. They should also be logical, respect the rules of evidence, be open to criticism, report methods and procedures, and make knowledge public. Explanations on how the natural world changes based on myths, personal beliefs, religious values, mystical inspiration, superstition, or authority may be personally useful and socially relevant, but they are not scientific.

Because all scientific ideas depend on experimental and observational confirmation, all scientific knowledge is, in principle, subject to change as new evidence becomes available. The core ideas of science such as the conservation of energy or the laws of motion have been subjected to a wide variety of confirmations and are therefore unlikely to change in the areas in which they have been tested.

In areas where data or understanding are incomplete, such as the details of human evolution or questions surrounding global warming, new data may well lead to changes in current ideas or resolve current conflicts.

In situations where information is still fragmentary, it is normal for scientific ideas to be incomplete, but this is also where the opportunity for making advances may be greatest.

Historical Perspectives

In history, diverse cultures have contributed sci-entific knowledge and technologic inventions.

Modern science began to evolve rapidly in Europe several hundred years ago. During the past two centuries, it has contributed significantly to the industrialization of Western and non-Western cul-tures. However, other, non-European cultures have developed scientific ideas and solved human problems through technology.

Usually, changes in science occur as small modi-fications in extant knowledge. The daily work of science and engineering results in incremental advances in our understanding of the world and our ability to meet human needs and aspirations. Much can be learned about the internal workings of science and the nature of science from study of individual scientists, their daily work, and their efforts to advance scientific knowledge in their area of study.

CONCLUSION

The material addressing evolution in the *National Science Education Standards* is embedded within the full range of content standards describing what students should know, understand, and be able to do in the natural sciences. Used in conjunction with standards for other parts of the science education system, the content standards—and their treatment of evolution—point toward the levels of scientific literacy needed to meet the challenges of the twenty-first century.

4

EVOLUTION AND THE NATURE OF SCIENCE

Teachers often face difficult questions about evolution, many from parents and others who object to evolution being taught. Science has good answers to these questions, answers that draw on the evidence supporting evolution and on the nature of science. This chapter presents short answers to some of the most commonly asked questions.

DEFINITIONS

What is Evolution?

Evolution in the broadest sense explains that what we see today is different from what existed in the past. Galaxies, stars, the solar system, and earth have changed through time, and so has life on earth.

Biological evolution concerns changes in living things during the history of life on earth. It explains that living things share common ancestors. Over time, evolutionary change gives rise to new species. Darwin called this process "descent with modification," and it remains a good definition of biological evolution today.

What is "creation science"?

The ideas of "creation science" derive from the conviction that God created the universe—including humans and other living things—all at once in the relatively recent past. However, scientists from many fields have examined these ideas and have found them to be scientifically insupportable. For example, evidence for a very young earth is incompatible with many different methods of establishing the age of rocks. Furthermore, because the basic proposals of creation science are not subject to test and verification, these ideas do not meet the criteria for science. Indeed, U.S. courts have ruled that ideas of creation science are religious views and cannot be taught when evolution is taught.

THE SUPPORTING EVIDENCE

How can evolution be scientific when no one was there to see it happen? This question reflects a narrow view of how science works. Things in science can be studied even if they cannot be directly observed or experimented on. Archaeologists study past cultures by examining the artifacts those cultures left behind. Geologists can describe past changes in sea level by studying the marks ocean waves left on rocks. Paleontologists study the fossilized remains of organisms that lived long ago.

Something that happened in the past is thus not "off limits" for scientific study. Hypotheses can be made about such phenomena, and these hypotheses can be tested and can lead to solid conclusions. Furthermore, many key aspects of evolution occur in relatively short periods that can be observed directly—such as the evolution in bacteria of resistance to antibiotics.

Isn't evolution just an inference?

No one saw the evolution of one-toed horses from three-toed horses, but that does not mean that we cannot be confident that horses evolved. Science is practiced in many ways besides direct observation and experimentation. Much scientific discovery is done through indirect experimentation and observation in which inferences are made, and hypotheses generated from those inferences are tested. For instance, particle physicists cannot directly observe subatomic particles because the particles are too small. They must make inferences about the weight, speed, and other properties of the particles based on other observations. A logical hypothesis might be something like this: If the weight of this particle is *Y*, when I bombard it, *X* will happen. If *X* does not happen, then the hypothesis is disproved. Thus, we can learn about the natural world even if we cannot directly observe a phenomenon —and that is true about the past, too.

In historical sciences like astronomy, geology, evolutionary biology, and archaeology, logical infer-ences are made and then tested against data. Sometimes the test cannot be made until new data are available, but a great deal has been done to help us understand the past.

For example, scorpionflies (*Mecoptera*) and true flies (*Diptera*) have enough similarities that entomologists consider them to be closely related. Scorpionflies have four wings of about the same size, and true flies have a large front pair of wings but the back pair is replaced by small club-shaped structures. If *Diptera* evolved from *Mecoptera*, as comparative anatomy suggests, scientists predicted that a fossil fly with four wings might be found—and in 1976 this is exactly what was discovered. Furthermore, geneticists have found that the number of wings in flies can be changed through mutations in a single gene.

Evolution is a well-supported theory drawn from a variety of sources of data, including observations about the fossil record, genetic information, the distribution of plants and animals, and the similarities across species of anatomy and development. Scientists have inferred that descent with modification offers the best scientific explanation for these observations.

Is evolution a fact or a theory?

The theory of evolution explains how life on earth has changed. In scientific terms, "theory" does not mean "guess" or "hunch" as it does in everyday usage. Scientific theories are explanations of natural phenomena built up logically from testable observations and hypotheses. Biological evolution is the best scientific explanation we have for the enormous range of observations about the living world. Scientists most often use the word "fact" to describe an observation. But scientists can also use fact to mean something that has been tested or observed so many times that there is no longer a compelling reason to keep testing or looking for examples. The occurrence of evolution in this sense is a fact. Scientists no longer question whether descent with modification occurred because the evidence supporting the idea is so strong.

Why isn't evolution called a law?

Laws are generalizations that *describe* phenomena, whereas theories *explain* phenomena. For example, the laws of thermodynamics describe what will happen under certain circumstances; thermodynamics theories explain why these events occur. Laws, like facts and theories, can change with better data. But theories do not develop into laws with the accumulation of evidence. Rather, theo-ries are the goal of science.

Don't many famous scientists reject evolution?

No. The scientific consensus around evolution is overwhelming. Those opposed to the teaching of evolution sometimes use quotations from prominent scientists out of context to claim that scientists do not support evolution. However, examination of the quotations reveals that the scientists are actually disputing some aspect of *how* evolution occurs, not *whether* evolution occurred. For example, the biologist Stephen Jay Gould once wrote that "the extreme rarity of transitional forms in the fossil record persists as the trade secret of paleontology." But Gould, an accomplished paleontologist and eloquent educator about evolution, was arguing about *how* evolution takes place. He was discussing whether the rate of change of species is constant and gradual or whether it takes place in bursts after long periods when little change occurs—an idea known as punctuated equilibrium. As Gould writes in response, "This quotation, although accurate as a partial citation, is dishonest in leaving out the following explanatory material showing my true purpose—to discuss rates of evolutionary change, not to deny the fact of evolution itself."

Gould defines punctuated equilibrium as follows:

Punctuated equilibrium is neither a creationist idea nor even a non-Darwinian evolutionary theory about sudden change that produces a new species all at once in a single generation.

Punctuated equilibrium accepts the conventional idea that new species form over hundreds or thousands of generations and through an extensive series of intermediate stages. But geological time is so long that even a few thousand years may appear as a mere "moment" relative to the several million years of existence for most species.

Thus, rates of evolution vary enormously and new species may appear to arise "suddenly" in geological time, even though the time involved would seem long, and the change very slow, when com-pared to a human lifetime.

Isn't the fossil record full of gaps?

Though significant gaps existed in the fossil record in the 19th century, many have been filled in. In addition, the consistent pattern of ancient to modern species found in the fossil record is strong evidence for evolution. The plants and animals living today are not like the plants and animals of the remote past. For example, dinosaurs were extinct long before humans walked the earth. We know this because no human remains have ever been found in rocks dated to the dinosaur era.

Some changes in populations might occur too rapidly to leave many transitional fossils. Also, many organisms were very unlikely to leave fossils, either because of their habitats or because they had no body parts that could easily be fossilized. However, in many cases, such as between primi-tive fish and amphibians, amphibians and reptiles, reptiles and mammals, and reptiles and birds, there are excellent transitional fossils.

Can evolution account for new species?

One argument sometimes made by supporters of "creation science" is that natural selection can produce minor changes within species, such as changes in color or beak size, but cannot generate new species from pre-existing species. However, evolutionary biologists have documented many cases in which new species have appeared in recent years (some of these cases are discussed in Chapter 2). Among most plants and animals, speciation is an extended process, and a single human observer can witness only a part of this process. Yet these observations of evolution at work provide powerful confirmation that evolution forms new species.

If humans evolved from apes, why are there still apes?

Humans did not evolve from modern apes, but humans and modern apes shared a common ancestor, a species that no longer exists. Because we shared a recent common ancestor with chimpanzees and gorillas, we have many anatomical, genetic, biochemical, and even behavioral similarities with the African great apes. We are less similar to the Asian apes—orangutans and gibbons—and even less similar to monkeys, because we shared common ancestors with these groups in the more distant past.

Evolution is a branching or splitting process in which populations split off from one another and gradually become different. As the two groups become isolated from each other, they stop sharing genes, and eventually genetic differences increase until members of the groups can no longer interbreed. At this point, they have become separate species. Through time, these two species might give rise to new species, and so on through millennia.

Doesn't the sudden appearance of all the "modern groups" of animals during the Cambrian explosion prove creationism?

During the Cambrian explosion, primitive representatives of the major phyla of invertebrate animals appeared—hard-shelled organisms like mollusks and arthropods. More modern representatives of these invertebrates appeared gradually through the Cambrian and the Ordovician periods.

"Modern groups" like terrestrial vertebrates and flowering plants were not present. It is not true that "all the modern groups of animals" appeared during this period.

Also, Cambrian fossils did not appear spontaneously. They had ancestors in the Precambrian period, but because these Precambrian forms were soft-bodied, they left fewer fossils. A characteristic of the Cambrian fossils is the evolution of hard body parts, which greatly improved the chance of fossilization. And even without fossils, we can infer relationships among organisms from biochemical information.

RELIGIOUS ISSUES

Can a person believe in God and still accept evolution?

Many do. Most religions of the world do not have any direct conflict with the idea of evolution. Within the Judeo-Christian religions, many people believe that God works through the process of evolution. That is, God has created both a world that is ever-changing and a mechanism through which crea-tures can adapt to environmental change over time. At the root of the apparent conflict between some religions and evolution is a misunderstanding of the critical difference between religious and scientific ways of knowing. Religions and science answer different questions about the world.

Whether there is a purpose to the universe or a purpose for human existence are not questions for science. Religious and scientific ways of knowing have played, and will continue to play, significant roles in human history.

No one way of knowing can provide all of the answers to the questions that humans ask. Consequently, many people, including many scien-tists, hold strong religious beliefs and simultaneously accept the occurrence of evolution.

Aren't scientific beliefs based on faith as well?

Usually "faith" refers to beliefs that are accepted without empirical evidence. Most religions have tenets of faith. Science differs from religion because it is the nature of science to test and retest explanations against the natural world. Thus, scientific explanations are likely to be built on and modified with new information and new ways of looking at old information. This is quite different from most religious beliefs.

Therefore, "belief" is not really an appropriate term to use in science, because testing is such an important part of this way of knowing. If there is a component of faith to science, it is the assumption that the universe operates according to regularities—for example, that the speed of light will not change tomorrow. Even the assumption of that regularity is often tested—and thus far has held up well. This "faith" is very different from religious faith. Science is a way of knowing about the natural world. It is limited to explaining the natural world through natural causes. Science can say nothing about the supernatural. Whether God exists or not is a question about which science is neutral.

LEGAL ISSUES

Why can't we teach creation science in my school?

The courts have ruled that "creation science" is actually a religious view. Because public schools must be religiously neutral under the U.S. Constitution, the courts have held that it is unconstitutional to present creation science as legitimate scholarship.

In particular, in a trial in which supporters of creation science testified in support of their view, a district court declared that creation science does not meet the tenets of science as scientists use the term (*McLean v. Arkansas Board of Education*). The Supreme Court has held that it is illegal to require that creation science be taught when evolution is taught (*Edwards v. Aguillard*). In addition, district courts have decided that individual teachers cannot advocate creation science on their own (*Peloza v. San Juan Capistrano School District* and *Webster v. New Lennox School District*).

Teachers' organizations such as the National Science Teachers Association, the National Association of Biology Teachers, the National Science Education Leadership Association, and many others also have rejected the science and pedagogy of creation science and have strongly discouraged its presentation in the public schools. (Statements from some of these organizations appear in Appendix C.) In addition, a coalition of religious and other organizations has noted in "A Joint Statement of Current Law" (see Appendix B) that "in science class, [schools] may present only genuinely scientific critiques of, or evidence for, any explanation of life on earth, but not religious critiques (beliefs unverifiable by scientific methodology)." Some argue that "fairness" demands the teaching of creationism along with evolution. But a science curriculum should cover science, not the religious views of particular groups or individuals.

EDUCATIONAL ISSUES

If evolution is taught in schools, shouldn't creationism be given equal time?

Some religious groups deny that microorganisms cause disease, but the science curriculum should not therefore be altered to reflect this belief. Most people agree that students should be exposed to the best possible scholarship in each field. That scholarship is evaluated by professionals and educators in those fields. In science, scientists as well as educators have concluded that evolution—and only evolution—should be taught in science classes because it is the only *scientific* explanation for why the universe is the way it is today.

Many people say that they want their children to be exposed to creationism in school, but there are thousands of different ideas about creation among the world's people. Comparative religions might comprise a worthwhile field of study but not one appropriate for a science class. Furthermore, the U.S. Constitution states that schools must be religiously neutral, so legally a teacher could not present any particular creationist view as being more "true" than others.

Why should teachers teach evolution when they already have so many things to teach and can cover biology without mentioning evolution?

Teachers face difficult choices in deciding what to teach in their limited time, but some ideas are of central importance in each discipline. In biology, evolution is such an idea. Biology is sometimes taught as a list of facts, but if evolution is introduced early in a class and in an uncomplicated manner, it can tie many disparate facts together. Most impor-tant, it offers a way to understand the astonishing complexity, diversity, and activity of the modern world. Why are there so many different types of organisms? What is the response of a species or community to a changing environment? Why is it so difficult to develop antibiotics and insecticides that are useful for more than a decade or two? All of these questions are easily discussed in terms of evolution but are difficult to answer otherwise.

A lack of instruction about evolution also can hamper students when they need that information to take other classes, apply for college or medical school, or make decisions that require a knowledge of evolution.

Should students be given lower grades for not believing in evolution?

No. Children's personal views should have no effect on their grades. Students are not under a compulsion to accept evolution. A grade reflects a teacher's assessment of a student's understanding. If a child does not understand the basic ideas of evolution, a grade could and should reflect that lack of understanding, because it is quite possible to comprehend things that are not believed.

Can evolution be taught in an inquiry-based fashion?

Any science topic can be taught in an inquiryoriented manner, and evolution is particularly amenable to this approach. At the core of inquiryoriented instruction is the provision for students to collect data (or be given data when collection is not possible) and to analyze the data to derive patterns, conclusions, and hypotheses, rather than just learning facts. Students can use many data sets from evolution (such as diagrams of anatomical differences in organisms) to derive patterns or draw connections between morphological forms and environmental conditions. They then can use their data sets to test their hypotheses. Students also can collect data in real time. For example, they can complete extended projects involving crossbreeding of fruit flies or plants to illustrate the genetic patterns of inheritance and the influence of the environment on survival. In this way, students can develop an understanding of evo-lution, scientific inquiry, and the nature of science.

ACTIVITIES FOR TEACHING

Prior chapters in this volume answer the what and why questions of teaching about evolution and the nature of science. As every educator knows, such discussions only set a stage. The actual play occurs when science teachers act on the basic content and wellreasoned arguments for inclusion of evolution and the nature of science in school science programs. This chapter goes beyond discussions of content and rationales. It presents, as examples of investigative teaching exercises, eight activities that science teachers can use as they begin developing students' understandings and abilities of evolution and the nature of science. The following descriptions briefly introduce each activity.

ACTIVITY 1: Introducing Inquiry and the Nature of Science

This activity introduces basic procedures involved in inquiry and concepts describing the nature of science. In the first portion of the activity the teacher uses a numbered cube to involve students in asking a question—what is on the unseen bottom of the cube?—and the students propose an explanation based on their observations.

Then the teacher presents the students with a second cube and asks them to use the available evidence to propose an explanation for what is on the bottom of this cube. Finally, students design a cube that they exchange and use for an evaluation. This activity provides students with opportunities to learn the abilities and understandings aligned with science as inquiry and the nature of science as described in the *National Science Education Standards*.1 Designed for grades 5 through 12, the activity requires a total of four class periods to complete. Lower grade levels might only complete the first cube and the evaluation where students design a problem based on the cube activity.

ACTIVITY 2: The Formulation of Explanations: An Invitation to Inquiry on Natural Selection

This activity uses the concept of natural selection to introduce the idea of formulating and testing a scientific hypothesis. Through a focused discussion approach, the teacher provides information and allows students time to think, interact with peers, and propose explanations for observations described by the teacher.

The teacher then provides more information, and the students continue their discussion based on the new information. This activity will help students in grades 5 through 8 develop abilities related to scientific inquiry and formulate understandings about the nature of science.

ACTIVITY 3: Investigating Natural Selection

In this activity, the students investigate one mechanism for evolution through a simulation that models the principles of natural selection and helps answer the question: How might biological change have occurred and been reinforced over time? The activity is designed for grades 9 through 12 and requires three class periods.

ACTIVITY 4: Investigating Common Descent: Formulating Explanations and Models

In this activity, students formulate explanations and models that simulate structural and biochemical data as they investigate the misconception that humans evolved from apes. The investigations require two 45-minute periods. They are designed for use in grades 9 through 12.

ACTIVITY 5: Proposing Explanations for Fossil Footprints

In this investigation, students observe and interpret "fossil footprint" evidence. From the evidence, they are asked to construct defensible hypotheses or explanations for events that took place in the geologic past. Estimated time requirements for this activity: two class periods. This activity is designed for grades 5 through 8.

ACTIVITY 6: Understanding Earth's Changes Over Time

Comparing the magnitude of geologic time to spans of time within a person's own lifetime is diffi-cult for many students. In this activity, students use a long paper strip and a reasonable scale to represent visually all of geologic time, including significant events in the development of life on earth as well as recent human events. The investigation requires two class periods and is appropriate for grades 5 through 12.

ACTIVITY 7: Proposing the Theory of Biological Evolution: Historical Perspective

This activity uses historical perspectives and the theme of evolution to introduce students to the nature of science. The teacher has students read short excerpts of original statements on evolution from Jean Lamarck, Charles Darwin, and Alfred Russel Wallace. These activities are intended as either supplements to other investigations or core activities. Designed for grades 9 through 12, the activities should be used as part of three class periods.

ACTIVITY 8: Connecting Population Growth and Biological Evolution

In this activity, students develop a model of the mathematical nature of population growth. The investigation provides an excellent opportunity for consideration of population growth of plant and animal species and the relationship to mechanisms promoting natural selection. This activity will require two class periods and is appropriate for grades 5 through 12.

The activities in this chapter do not represent a curriculum. They are directed, instead, toward other purposes.

First, they present examples of standards-based instructional materials. In this case, the level of organization is an activity—one to five days of lessons—and not a larger level of organization such as a unit of several weeks, a semester, or a year. Also, these exercises generally do not use biological materials, such as fruit flies, or computer simulations. The use of these instructional materials in the curriculum greatly expands the range of possible investigations.

Second, these activities demonstrate how existing exercises can be recast to emphasize the importance of inquiry and the fundamental concepts of evolution. Each of these exercises was derived from already existing activities that were revised to reflect the *National Science Education Standards*.

For each exercise, student outcomes drawn from the *Standards* are listed to focus attention on the concepts and abilities that students are meant to develop.

Third, the activities demonstrate some, but not all, of the criteria for curricula to be described in Chapter 7. For example, several of the activities emphasize inquiry and the nature of science while others focus on concepts related to evolution. All activities use an instructional model, described in the next section, that increases coherence and enhances learning.

Finally, there remains a paucity of instructional materials for teaching evolution and the nature of science. Science teachers who recognize this need are encouraged to develop new materials and lessons to introduce the themes of evolution and the nature of science. (See http://www4.nas.edu/ opus/evolve.nsf)

DEVELOPING STUDENTS' UNDERSTANDING AND ABILITIES: THE CURRICULUM PERSPECTIVE

For students to develop an understanding of evolution and the nature of science requires many years and a variety of educational experiences. Teachers cannot rely on single lessons, chapters, or biology and earth science courses for students to integrate the ideas presented in this document into their own understanding. In early grades (K–4) students might learn the fundamental concepts associated with "characteristics of organisms," "life cycles," and "organisms and environments." In middle grades they learn more about "reproduction and heredity" and "diversity and adaptation of organisms." Such learning experiences, as described in the *National Science Education Standards*, set a firm foundation for the study of biological evolution in grades 9–12.

The slow and steady development of concepts such as evolution and related ideas such as natural selection and common descent requires careful consideration of the overall structure and sequence of learning experiences. Although this chapter does not propose a curriculum or a curriculum framework, current efforts by Project 2061 of the American Association for the Advancement of Science (AAAS) demonstrate the interrelated nature of students' understanding of science concepts and emphasize the importance of welldesigned curricula at several levels of organization (for example, activities, units, and school science programs). The figure on the next page presents the "Growth-of-Understanding Map for Evolution and Natural Selection" based on *Benchmarks for Science Literacy*.2

DEVELOPING STUDENT UNDERSTANDING AND ABILITIES: THE INSTRUCTIONAL PERSPECTIVE

The activities in the chapter incorporate an instructional model, summarized in the accompanying box, that includes five steps: engagement, exploration, explanation, elaboration, and evaluation. Just as scientific investigations originate with a question that engages a scientist, so too must students engage in the activities of learning. The activities therefore begin with a strategic question that gets students thinking about the content of the lesson.

Once engaged, students need time to explore ideas before concepts begin to make sense. In this exploration phase, students try their ideas, ask questions, and look for possible answers to questions. Students use inquiry strategies; they try to relate their ideas to those of other students and to what scientists already know about evolution. In the third step, students can propose answers and develop hypotheses. Also in this step, the teacher explains what scientists know about the questions. This is the step when teachers should make the major concepts explicit and clear to the students.

Educators understand that informing students about a concept does not necessarily result in their immediate comprehension and understanding of the idea. These activities therefore provide a step referred to as elaboration in which students have opportunities to apply their ideas in new and slightly different situations.

Finally, how well do students understand the concepts, or how successful are they at applying the desired skills? These are the questions to be answered during the evaluation phase. Ideally, evaluations are more than tests. Students should have opportunities to see if their ideas can be applied in new situations and to compare their understanding with scientific explanations of the same phenomena.

ACTIVITY 1: INTRODUCING INQUIRY AND THE NATURE OF SCIENCE

This activity introduces basic procedures involved in inquiry and concepts describing the nature of science. In the first portion of the activity the teacher uses a numbered cube to involve students in asking a question—what is on the bottom?— and the students propose an explanation based on their observations. Then the teacher presents the students with a second cube and asks them to use the available evidence to propose an explanation for what is on the bottom of this cube.

Finally, students design a cube that they exchange and use for an evaluation. This activity provides students with opportunities to learn the abilities and understandings aligned with science as inquiry and the nature of science as described in the *National Science Education Standards.* Designed for grades 5 through 12, the activity requires a total of four class periods to complete. Lower grade levels might only complete the first cube and the evaluation where students design a problem based on the cube activity.

Standards-Based Outcomes

This activity provides all students with opportunities to develop abilities of scientific inquiry as described in the *National Science Education Standards.* Specifically, it enables them to:

- identify questions that can be answered through scientific investigations,
- design and conduct a scientific investigation,
- use appropriate tools and techniques to gather, analyze, and interpret data,
- develop descriptions, explanations, predictions, and models using evidence,
- think critically and logically to make relationships between evidence and explanations,
- recognize and analyze alternative explanations and predictions, and
- communicate scientific procedures and explanations.

This activity also provides all students opportunities to develop understanding about inquiry and the nature of science as described in the *National Science Education Standards*. Specifically, it introduces the following concepts:

- Different kinds of questions suggest different kinds of scientific investigations.
- Current scientific knowledge and understanding guide scientific investigations.
- Technology used to gather data enhances accuracy and allows scientists to analyze and quantify results of investigations.
- Scientific explanations emphasize evidence, have logically consistent arguments, and use scientific principles, models, and theories.
- Science distinguishes itself from other ways of knowing and from other bodies of knowledge through the use of empirical standards, logical arguments, and skepticism, as scientists strive for the best possible explanations about the natural world.

Science Background for Teachers

The pursuit of scientific explanations often begins with a question about a natural phenomenon. Science is a way of developing answers, or improving explanations, for observations or events in the natural world. The scientific question can emerge from a child's curiosity about where the dinosaurs went or why the sky is blue. Or the question can extend scientists' inquiries into the process of extinction or the chemistry of ozone depletion.

Once the question is asked, a process of scientific inquiry begins, and there eventually may be an answer or a proposed explanation. Critical aspects of science include curiosity and the freedom to pursue that curiosity. Other attitudes and habits of mind that characterize scientific inquiry and the activities of scientists include intelligence, honesty, skepticism, tolerance for ambiguity, openness to new knowledge, and the willingness to share knowledge publicly.

Scientific inquiry includes systematic approaches to observing, collecting information, identifying significant variables, formulating and testing hypotheses, and taking precise, accurate, and reliable measurements.

Understanding and designing experiments are also part of the inquiry process. Scientific explanations are more than the results of collecting and organizing data. Scientists also engage in important processes such as constructing laws, elaborating models, and developing hypotheses based on data.

These processes extend, clarify, and unite the observations and data and, very importantly, develop deeper and broader explana-tions. Examples include the taxonomy of organisms, the periodic table of the elements, and theories of common descent and natural selection.

One characteristic of science is that many explanations continually change. Two types of changes occur in scientific explanations: new explanations are developed, and old explanations are modified.

Just because someone asks a question about an object, organism, or event in nature does not necessarily mean that person is pursuing a scientific explanation. Among the conditions that must be met to make explanations scientific are the following:

- *Scientific explanations are based on empirical observations or experiments.* The appeal to authority as a valid explanation does not meet the requirements of science. Observations are based on sense experiences or on an extension of the senses through technology.
- *Scientific explanations are made public.* Scientists make presentations at scientific meetings or publish in professional journals, making knowledge public and available to other scientists.
- *Scientific explanations are tentative.* Explanations can and do change. There are no scientific truths in an absolute sense.
- *Scientific explanations are historical.* Past explanations are the basis for contemporary explanations, and those, in turn, are the basis for future explanations.

- *Scientific explanations are probabilistic.* The statistical view of nature is evident implicitly or explicitly when stating scientific predictions of phenomena or explaining the likelihood of events in actual situations.
- *Scientific explanations assume cause-effect relationships.* Much of science is directed toward determining causal relationships and developing explanations for interactions and linkages between objects, organisms, and events. Distinctions among causality, correlation, coincidence, and con-tingency separate science from pseudoscience.
- *Scientific explanations are limited.* Scientific explanations sometimes are limited by technology, for example, the resolving power of microscopes and telescopes. New technologies can result in new fields of inquiry or extend current areas of study. The interactions between technology and advances in molecular biology and the role of technology in planetary explorations serve as examples. Science cannot answer all questions. Some questions are simply beyond the parameters of science.

Many questions involving the meaning of life, ethics, and theology are examples of questions that science cannot answer. Refer to the *National Science Education Standards* for Science as Inquiry (pages 145-148 for grades 5-8 and pages 175-176 for grades 9-12), History and Nature of Science Standards (pages 170-171 for grades 5-8 and pages 200-204 for grades 9-12), and Unifying Concepts and Processes (pages 116-118). Chapter 3 of this document also contains a discussion of the nature of science.

Materials and Equipment

- 1 cube for each group of four students (blackline masters are provided).

(Note: you may wish to complete the first portion of the activity as a demonstration for the class. If so, construct one large cube using a cardboard box. The sides should have the same numbers and markings as the black-line master.)

- 10 small probes such as tongue depressors or pencils.

- 10 small pocket mirrors.

Instructional Strategy

Engage Begin by asking the class to tell you what they know about how scientists do their work. How would they describe a scientific investigation? Get students thinking about the process of scientific inquiry and the nature of science. This is also an opportunity for you to assess their current understanding of science. Accept student answers and record key ideas on the overhead or chalkboard.

Explore (The first cube activity can be done as a demonstration if you construct a large cube and place it in the center of the room.) First, have the students form groups of three or four. Place the cubes in the center of the table where the students are working. The students should not touch, turn, lift, or open the cube. Tell the students they have to identify a question associated with the cube.

Allow the students to state their questions. Likely questions include:

- What is in the cube?
- What is on the bottom of the cube?
- What number is on the bottom?

You should direct students to the general question, *what is on the bottom of the cube*? Tell the students that they will have to answer the question by proposing an explanation, and that they will have to convince you and other students that their answer is *based on evidence*. (Evidence refers to observations the group can make about the visible sides of the cube.) Allow the students time to explore the cube and to develop answers to their question. Some observations or statements of fact that the students may make include:

- The cube has six sides.
- The cube has five exposed sides.
- The numbers and dots are black.

- The exposed sides have numbers 1, 3, 4, 5, and 6.
- The opposite sides add up to seven.
- The even-numbered sides are shaded.
- The odd-numbered sides are white.

Ask the students to use their observations (the data) to propose an answer to the question: *What is on the bottom of the cube*? The student groups should be able to make a statement such as: *We conclude there is a 2 on the bottom*. Students should present their reasoning for this conclusion.

For example, they might base their conclusion on the observation that the exposed sides are 1, 3, 4, 5, and 6, and because 2 is missing from the sequence, they conclude it is on the bottom. Use this opportunity to have the students develop the idea that combining two different but logically related observations creates a stronger explanation. For example, 2 is missing in the sequence (that is, 1, _, 3, 4, 5, 6) and that opposite sides add up to 7 (that is, 1—6; 3—4; _—5) and because 5 is on top, and 5 and 2 equal 7, 2 could be on the bottom.

If done as a demonstration, you might put the cube away without showing the bottom or allowing students to dismantle it. Explain that scientists often are uncertain about their proposed answers, and often have no way of knowing the absolute answer to a scientific question. Examples such as the exact ages of stars and the reasons for the extinction of prehistoric organisms will support the point.

Explain Begin the class period with an explanation of how the activity simulates scientific inquiry and provides a model for science. Structure the discussion so students make the connections between their experiences with the cube and the key points (understandings) you wish to develop. Key points from the *Standards* include the following:

- Science originates in questions about the world.
- Science uses observations to construct explanations (answers to the questions). The more observations you

had that supported your proposed explanation, the stronger your explanation, even if you could not confirm the answer by examining the bottom of the cube.

- Scientists make their explanations public through presentations at professional meetings and journals.
- Scientists present their explanations and critique the explanations proposed by other scientists. The activity does not explicitly describe "the scientific method." The students had to work to answer the question and probably did it in a less than systematic way. Identifiable elements of a method—such as observation, data, and hypotheses—were clear but not applied systematically.

You can use the experiences to point out and clarify scientific uses of terms such as observation, hypotheses, and data.

For the remainder of the second class period you should introduce the "story" of an actual scientific discovery. Historic examples such as Charles Darwin would be ideal. You could also assign stu-dents to prepare brief reports that they present.

Elaborate The main purpose of the second cube is to extend the concepts and skills introduced in the earlier activities and to introduce the role of prediction, experiment, and the use of technology in scientific inquiry.

The problem is the same as the first cube: *What is on the bottom of the cube?* Divide the class into groups of three and instruct them to make observations and propose an answer about the bottom of the cube. Student groups should record their factual statements about the second cube. Let students identify and organize their observations. If the students are becoming too frustrated, provide helpful suggestions. Essential data from the cube include the following (see black-line master):

- Names and numbers are in black.
- Exposed sides have either a male or female name.
- Opposing sides have a male name on one side and a female name on the other.
- Names on opposite sides begin with the same letters.

- The number in the upper-right corner of each side corresponds to the number of letters in the name on that side.
- The number in the lower-left corner of each side corresponds to the number of the first letter that the names on opposite sides have in common.
- The number of letters in the names on the five exposed sides progresses from three (Rob) to seven (Roberta).

Four names, all female, could be on the bottom of the cube: Fran, Frances, Francene, and Francine. Because there are no data to show the exact name, groups might have different hypotheses. Tell the student groups that scientists use pat-terns in data to make predictions and then design an experiment to assess the accuracy of their prediction. This process also produces new data.

Tell groups to use their observations (the data) to make a prediction of the number in the upper-right corner of the bottom. The predictions will most likely be 4, 7, or 8. Have the team decide which corner of the bottom they wish to inspect and why they wish to inspect it.

The students might find it difficult to determine which corner they should inspect. Let them struggle with this and even make a mistake—this is part of science! Have one student obtain a utensil, such as a tweezers, probe, or tongue depressor, and a mirror. The student may lift the designated corner less than one inch and use the mirror to look under the corner. This simulates the use of tech-nology in a scientific investigation. The groups should describe the data they gained by the "experiment." Note that the students used technology to expand their observations and understanding about the cube, even if they did not identify the corner that revealed the most productive evidence.

If students observe the corner with the most productive information, they will discover an 8 on the bottom. This observation will confirm or refute the students' working hypotheses. Francine or Francene are the two possible names on the bottom. The students propose their answer to the question and design another experiment to answer the question. Put the cube away without revealing the bottom. Have each of the student groups present brief reports on their investigation.

Evaluate The final cube is an evaluation. There are two parts to the evaluation. First, in groups of three, students must create a cube that will be used as the evaluation exercise for other groups. After a class period to develop a cube, the student groups should exchange cubes. The groups should address the same question: *What is on the bottom of the cube?* They should follow the same rules—for example, they cannot pick up the cube. The groups should prepare a written report on the cube developed by their peers. (You may have the students present oral reports using the same format.) The report should include the following:

- title,
- the question they pursued,
- observation—data,
- experiment—new data,
- proposed answer and supporting data,
- a diagram of the bottom of the cube, and
- suggested additional experiments.

Due to the multiple sources of data (information), this cube may be difficult for students. It may take more than one class period, and you may have to provide resources or help with some information. Remember that this activity is an evaluation. You may give some helpful hints, especially for information, but since the evaluation is for inquiry and the nature of science you should limit the information you provide on those topics. Student groups should complete and hand in their reports.

If student groups cannot agree, you may wish to make provisions for individual or "minority reports." You may wish to have groups present oral reports (a scientific conference). You have two opportunities to evaluate students on this activity: you can evaluate their understanding of inquiry and the nature of science as they design a cube, and you can assess their abilities and understandings as they figure out the unknown cube.

ACTIVITY 2: THE FORMULATION OF EXPLANATIONS: AN INVITATION TO INQUIRY ON NATURAL SELECTION

This activity uses the concept of natural selection to introduce the idea of formulating and testing scientific hypotheses. Through a focused discussion approach, the teacher provides information and allows students time to think, interact with peers, and propose explanations for observations described by the teacher. The teacher then pro-vides more information, and the students continue their discussion based on the new information. This activity will help students in grades 5 through 8 develop several abilities related to scientific inquiry and formulate understandings about the nature of science as presented in the *National Science Education Standards.*

This activity is adapted with permission from *BSCS: Biology Teachers' Handbook.*3

Standards-Based Outcomes

This activity provides all students with opportunities to develop the abilities of scientific inquiry as described in the *National Science Education Standards.* Specifically, it enables them to:

- identify questions that can be answered through scientific investigations,
- design and conduct a scientific investigation,
- use appropriate tools and techniques to gather, analyze, and interpret data,
- develop descriptions, explanations, predictions, and models using evidence,
- think critically and logically to make relationships between evidence and explanations,
- recognize and analyze alternative explanations and predictions, and
- communicate scientific procedures and explanations.

This activity also provides all students opportunities to develop understandings about inquiry, the nature of science, and biological evolution as described in the *National Science Education Standards*.

Specifically, it conveys the following concepts:

- Different kinds of questions suggest different kinds of scientific investigations.
- Current scientific knowledge and understanding guide scientific investigations.
- Technology used to gather data enhances accuracy and allows scientists to analyze and quantify results of investigations.
- Scientific explanations emphasize evidence, have logically consistent arguments, and use scientific principles, models, and theories.
- Species evolve over time. Evolution is the consequence of the interactions of (1) the potential for a species to increase its numbers, (2) the genetic variability of offspring due to mutation and recombination of genes, (3) a finite supply of the resources required for life, and (4) the ensuing selection of those offspring better able to survive and leave offspring in a particular environment.

Science Background for Teachers

Many biological theories can be thought of as developing in five interrelated and overlapping stages. The first is a period of extensive observation of nature or analyzing the results of experiments. Darwin's observations would be an exam-ple of the former. Second, these observations lead scientists to ponder questions of "how" and "why." In the course of answering these questions, scientists infer explanations or make conjectures as working hypotheses.

Third, in most cases, scientists submit hypotheses to formal, rigorous tests to check the validity of the hypotheses. At this point the hypotheses can be confirmed, falsified and rejected (not supported with evidence), or modified based on the evidence. This is a stage of experimentation. Fourth, scientists propose formal explanations by making public presentations at professional meetings or publishing their results in peer-reviewed journals. Finally, adoption of an explanation is recognized by other scientists as they begin referring to and using the explanation in their research and publications.

This activity focuses on the second and third stages in this brief summary of the development of biological theories. Chapters 2 and 3 of this document provide further discussion of these points. Review the "History and Nature of Science" and "Science as Inquiry" sections of the *National Science Education Standards* for further background on scientific investigations.

Materials and Equipment

None required.

Instructional Strategy

Engage Have the students work in groups of two or three. Begin by engaging the students with the problem and the basic information they will need to formulate a hypothesis.

TO THE STUDENTS: A farmer was working with dairy cattle at an agricultural experiment station. The population of flies in the barn where the cattle lived was so large that the animals' health was affected. So the farmer sprayed the barn and the cattle with a solution of insecticide A. The insecticide killed nearly all the flies.

Sometime later, however, the number of flies was again large. The farmer again sprayed with the insecticide. The result was similar to that of the first spraying. Most, but not all, of the flies were killed.

Again within a short time the population of flies increased, and they were again sprayed with the insecticide. This sequence of events was repeated five times; then it became apparent that insecticide A was becoming less and less effective in killing the flies.

Explore Imagine that the farmer consulted a group of student researchers. Have the student groups discuss the problem and prepare several different hypotheses to account for the observations.

They should share their results with the class. Students might propose explanations similar to the following:

1. Decomposition of insecticide A with age.
2. The insecticide is effective only under certain environmental conditions—for example, certain temperatures and levels of humidity—which changed in the course of the work.
3. The flies genetically most susceptible to the insecticide were selectively killed. (This item should not be elicited at this point or developed if suggested.)

TO THE STUDENTS: One farmer noted that one large batch of the insecticide solution had been made and used in all the sprayings. Therefore, he suggested the possibility that the insecticide solution decomposed with age. Have the student groups suggest at least two different approaches to test this hypothesis. The students may propose that investigation of several different predictions of a hypothesis contributes to the reliability of the conclusions drawn. In the present instance, one approach would be to use sprays of different ages on different populations of flies. A quite different approach would consist simply of making a chemical analysis of fresh and old solutions to determine if changes had occurred.

TO THE STUDENTS: The student researchers made a fresh batch of insecticide A. They used it instead of the old batch on the renewed fly population at the farmer's barn. Nevertheless, despite the freshness of the solution, only a few of the flies died.

The same batch of the insecticide was then tried on a fly population at another barn several miles away. In this case, the results were like those originally seen at the experiment station—that is, most of the flies were killed. Here were two quite different results with a fresh batch of insecticide. Moreover, the weather conditions at the time of the effective spraying of the distant barn were the same as when the spray was used without success at the experiment station. Stop and have the student groups analyze the observations and list the major components of the problem and subsequent hypotheses. They might list what they know, what they propose as explana-tions, and what they could do to test their explanations. Students might identify the following:

1. Something about the insecticide.
2. The conditions under which the insecticide was used.
3. The way in which the insecticide was used.
4. The organisms on which the insecticide was used.

TO THE STUDENTS: So far our hypotheses have had to do with just a few of these components. Which ones?

The hypotheses so far have concerned only "something about the insecticide" and "the condition under which the insecticide was used," items 1 and 2 above.

TO THE STUDENTS: The advantage of analyzing a problem, as we have done in our list, consists in the fact that it lets us see what possibilities we have not considered. What possibilities in the list have we not considered in forming our hypotheses?

Item 3, "the way in which the insecticide was used," may be pursued as a further exercise if the teacher so wishes. However, emphasis should be placed on Item 4, "the organisms on which the insecticide was used." This item is developed next.

Explain TO THE STUDENTS: Let us see if we can investigate the interactions between insecticide A and the flies. From your knowledge of biology, think of something that might have happened within the fly population that would account for the decreasing effectiveness of insecticide A. The students may need help here, even if they have learned something about evolution and natural selection. Here is one way to help:

Ask the students to remember that after the first spraying, most, *but not all,* of the flies were killed. Ask them where the new population of flies came from—that is, who were the parents of the next generation of flies? Were the parents among the flies more susceptible or more resistant to the effects of insecticide A?

Then remind them that the barn was sprayed again. If there are differences in the population to insecticide A susceptibility, which individuals would be more likely to survive this spraying? Remind them that dead flies do not produce offspring—only living ones can. The students might thus be led to see that natural selection, in this case in an imposed environment (the presence of the insecticide), might have resulted in the survival of only those individuals that were best adapted to live in the new environment (one with the insecticide). Because this activity centers on the formulation of explanations, it is important to introduce students to the scientific process they have been using. Following is a discussion from the *National Science Education Standards* that can serve as the basis for the explanation phase of the activity.

Evidence, Models, and Explanation

> *Evidence consists of observations and data on which to base scientific explanations. Using evidence to understand interactions allows individuals to predict changes in natural and designed systems.*
>
> *Models are tentative schemes or structures that correspond to real objects, events, or classes of events, and that have explanatory power. Models help scientists and engineers understand how things work. Models take many forms, including physical objects, plans, mental constructs, mathematical equations, and computer simulations.*
>
> *Scientific explanations incorporate existing scientific knowledge and new evidence from observations, experiments, or models into internally consistent, logical statements. Different terms, such as "hypothesis," "model," "law," "principle," "theory," and "paradigm," are used to describe various types of scientific explanations. As students develop and as they understand more science concepts and processes, their explanations should become more sophisticated. That is, their scientific explanations should more frequently include a rich scientific knowledge base, evidence of logic, higher levels of analysis, greater tolerance of criticism and uncertainty, and a clearer demonstration of the relationship between logic, evidence, and current knowledge.*

Elaborate Give the students a new problem— for example one of the investigations from *The Beak of the Finch*5 or *Darwin's Dreampond.*6 Have them work in groups to propose an explanation. The students should emphasize the role of hypotheses in the development of scientific explanations.

Evaluate Have the students consider the following case. Suppose a group of farmers notices the gradual acquisition of resistance to insecticide A over a period of months.

They locate two other equally powerful although chemically unrelated insecticides, insecticides B and C.

The local Agriculture Department sets up a program whereby all the farmers in the state will use only insecticide A for the current year. No one is to use insecticides B or C.

The following year, everyone is directed to use insecticide B rather than insecticide A. The fly population, which had become resistant to insecticide A, is now susceptible to insecticide B and can be kept under control much more thoroughly than if the farmers had continued using insecticide A.

At the beginning of the third year, all of the farmers begin using insecticide C, which again reduces the fly population greatly. As the fourth year begins, insecticide A is again used, and it proves to once again be extremely effective against the flies. Have students analyze this situation and propose an explanation of what has happened. How would they design an investigation to support or reject their hypothesis?

ACTIVITY 3: INVESTIGATING NATURAL SELECTION

In this activity, the students experience one mechanism for evolution through a simulation that models the principles of natural selection and helps answer the question: How might biological change have occurred and been reinforced over time? The activity is designed for grades 9 through 12 and requires three class periods. This activity is adapted with permission from *BSCS Biology: A Human Approach.*

Standards-Based Outcomes

This activity provides all students opportunities to develop understandings of biological evolution as described in the *National Science Education Standards*. Specifically, it conveys the concepts that:

- Species evolve over time. Evolution is the consequence of the interaction of (1) the potential for a species to increase in number, (2) the genetic vari-ability of offspring due to mutation and recombination of genes, (3) a finite supply of the resources required for life, and (4) the ensuing selection of those offspring better able to survive and leave offspring in a particular environment. Item 4 is the primary emphasis of this activity. Teachers can introduce the other factors as appropriate.
- Natural selection and its evolutionary consequences provide a scientific explanation for the fossil record of ancient life forms, as well as for the striking molecular similarities observed among the diverse species of living organisms.
- Some living organisms have the capacity to produce populations of almost infinite size, but environments and resources are finite. The fundamental tension has profound effects on the interactions among organisms.

Science Background for Teachers

Many students have difficulty with the fundamental concepts of evolution. For example, some students express misconceptions about natural selection because they do not understand the relationship between variations within a population and the potential effect of those variations as the population continues to grow in numbers in an environment with limited resources. This is a dynamic understanding that derives from the four ideas presented in the learning outcomes for this activity.

This activity emphasizes natural selection. In particular, it presents students with the predator-prey relationship as one example of how natural selection operates in nature.

Students should understand that the process of evolution has two steps, referred to as genetic variation and natural selection. The first step is the development of genetic variation through changes such as genetic recombination, gene flow, and mutations. The second step, and the point of this activity, is selection. Differential survival and reproduction of organisms is due to a variety of environmental factors such as predator-prey relationships, resource shortages, and change of habitat. In any generation only a small percentage of organisms survives. Survival depends on an organism's genetic constitution that will, given circum-stances such as limited resources, give a greater probability of survival and reproduction.8

Materials and Preparation (per class of 32)

8 petri dish halves

8 36-x 44-in. pieces of fabric, 4 each of 2 different patterns

8 sheets of graph paper

8 zip-type plastic sandwich bags containing 120 paper dots, 20 each of 6 colors (labeled "Beginning Population")

8 sets of colored pencils with colors similar to the paper dot colors

8 zip-type plastic sandwich bags of spare paper dots in all colors watch or clock with a second hand computer with spreadsheet software program (optional)

24 forceps (optional)

Choose fabric patterns that simulate natural environments, such as floral, leaf, or fruit prints. The patterns should have several colors and be of intricate design; small prints work better than large blocky prints. Select two designs, each with a different predominant color. Label one design Fabric A and the other Fabric B. The use of two designs enables the students to demonstrate the evolution of different color types from the same starting population.

Use a paper punch to punch out quarter-inch paper dots from construction paper of six different colors. Select two light colors (including white) and two dark colors so that they will compete against each other. Include at least two colors that blend well with the fabrics. For each color, put 100 dots into each of 8 zip-type plastic sandwich bags. Put 20 dots of each color (for a total of 120 dots of 6 colors) into each of 24 additional bags. Label these bags "Beginning Population." Enlist student aides or ask for student volunteers to punch dots or stuff bags at home or after school. As an alternative to paper dots, you might try colored aquarium gravel or colored rice. Both are heavier than paper dots and are less likely to blow around the room. You could color the rice grains with food dyes according to the criteria specified above for the dots. You also might use gift-wrap-ping paper instead of the pieces of fabric.

Instructional Strategy

Engage Begin by asking students what they know about the theory of natural selection. Ask them what predator-prey relationships have to do with biological evolution. Use the discussion as a means to have them explain how they think evolution occurs and the role of predator-prey relations in the process. At this point in the lesson, accept the variety of student responses, and determine any misconceptions the students express. You could present a historical example (see the discussion of fossils in chapter 3 of this volume) or an example from *The Beak of the Finch* by Jonathan Weiner or *Darwin's Dreampond* by Tijs Goldschmidt.

Because the instructional procedures are complex for this activity, you will have to be fairly explic-it about the process. Tell the students they will work in teams of four. (If your class does not divide evenly, use teams of five). The activity calls for half of the teams to use Fabric A and half of the teams to use Fabric B. It will help if you go through a "trial run" before students begin the activity.

Explore Step 1. Tell the students to pick a "game warden" from each group of four. The other group members will be the predators. Step 2. Examine the paper dots in the bags labeled "Starting Population" and record the number of individuals (dots) of each color. All of the dots represent individuals of a particular species, and the individuals can be one of six colors. Step 3. Make certain that half of the teams use Fabric A and half use Fabric B. The procedures remain the same for both groups.

Steps 4 and 5. Tell the predators to turn away from the habitats. The game warden then spreads one of the bags of "Beginning Population" across the fabric and tells the predators to turn around and gather prey—i.e., the dots. The predators must stop hunting (picking up dots) when the game warden says "Stop" in 20 seconds. If the predators have difficulty picking up the paper dots, provide forceps.

Step 6. After the hunting is stopped, the students should carefully collect all of the dots that remain on the fabric and sort them by color. The game wardens are responsible for recording these data on the graph paper using the colored pencils corresponding to the dot colors.

Step 7. To simulate reproduction among the paper dots, add three paper dots for each remaining dot of that color. These paper dots, obtained from the bags containing extra dots, represent offspring.

Step 8. Repeat the predation using the second generation of dots. Again record the number of remaining dots in the second generation.

Step 9. Explain to the students that they do not have to simulate reproduction as they did before, but rather should calculate the number of individuals that would be in the third-generation beginning population.

Step 10. The construction and analysis of bar graphs is a critical and time-consuming part of this activity. Place the color of survivors on the hori-zontal axis and the number of the beginning population (or second generation) on the vertical axis of this activity.

If you have ready access to computers and spreadsheet programs, you could incorporate the use of spreadsheets during this step.

Explain Step 11. Study the bar graphs of each generation. Discuss the following questions (possible student responses are included).

- Which, if any, colors of paper dots survived better than others in the second-and third-generation beginning populations of paper dots? Answers will vary depending on the color of the fabric that the students used. The beginning populations for the second and third generations should include more dots that are of colors similar to the fabric and fewer dots that are of colors that stand out against the fabric. The change between the first and third generations should be more dramatic than the change between the first and second generations.
- What might be the reason that predators did not select these colors as much as they did other colors? Some colors were better camouflaged than others—they blended into the environment.
- What effect did capturing a particular color dot have on the numbers of that color in the following generations?

When an individual is removed from a population and dies, in this case through predation, that individual no longer reproduces. The students should realize that heavy predation leads to a decrease in the size of the population and in the size of the gene pool.

Step 12. Allow the students enough time to resort the colored dots into the appropriate bags. Be sure the students recount the dots in each bag and replace missing dots. Have a three-hole punch and construction paper on hand to replace lost dots.

Elaborate This portion of the activity provides you with an opportunity to assess the learners' understanding of evolution and the mechanisms by which it occurs. Before the students begin to work on these tasks, display a piece of Fabric A and a piece of Fabric B and ask the learners to post their third generation bar graphs beside the fabric that they used. The learners now will benefit by comparing their results with those from other teams that used the same fabric as well as with those from teams that used a different fabric. These comparisons will give them more data with which to construct explanations for the results that they see.

1. How well do the class data support your team's conclusions in Step 11?

   Students need to be able to analyze the relationship between their response in Step 11 and the cumulative data. The specific response should address the relationship between the team data and the class data.

2. Imagine a real-life predator-prey relationship and write a paragraph that describes how one or more characteristics of the predator population or the prey population might change as a result of natural selection.

The students should explain that variation exists in populations. Individuals with certain characteristics are better adapted than other individuals to their environment, and consequently survive to produce offspring; less well-adapted individuals do not. The offspring, in turn, possess characteristics similar to those of their parents, and that makes them better adapted to the environment as well. These two concepts are the basis of natural selection, and they explain how populations evolve. Little variation in a population of organisms would mean that fewer differences would be expressed in the offspring. Fewer differences would mean that individuals would have similar advantages and disadvantages in the prevailing environmental conditions. This similarity, in turn, would mean that their survival and reproductive rates would be similar, so few heritable differences then would be passed on to the next generation.

Evaluate Have the students write one paragraph that summarizes their understanding of biological evolution. Refer to the learning outcomes and the *National Science Education Standards*. Expect that students will describe that in a population of organisms, variation exists among characteristics that parents pass on to their offspring. Individuals with certain characteristics might have a slight advantage over other individuals and thus live longer and reproduce more. If this advantage remains, the difference would be more noticeable over time. These changes could eventually lead to new species. The process of natural selection, then, provides an explanation for the relatedness of organisms and for biological change across time.

ACTIVITY 4: INVESTIGATING COMMON DESCENT: FORMULATING EXPLANATIONS AND MODELS

In this activity, students formulate explanations and models that simulate structural and biochemical data as they investigate the misconception that humans evolved from apes. The activities require two 45-minute periods. They are designed for use in grades 9 through 12. This activity is adapted with permission from *Evolution: Inquiries into Biology and Earth Science* by *BSCS*.9

Standards-Based Outcomes

This activity provides opportunities for all students to develop abilities of scientific inquiry as described in the *National Science Education Standards.* Specifically, it enables them to:

- formulate descriptions, explanations, predictions, and models using evidence,
- think critically and logically to make relationships between evidence and explanations, and
- recognize and analyze alternative explanations and predictions.

In addition, the activity provides all students opportunities to develop fundamental understandings in the life sciences as described in the National Science Education Standards.

Specifically, it conveys the following concepts:

- In all organisms, the instructions for specifying the characteristics of the organism are carried in DNA, a large polymer formed from subunits of four kinds (A, G, C, and T). The chemical and structural properties of DNA explain how the genetic information that underlies heredity is both encoded in genes (as a string of molecular "letters") and replicated (by a templating mechanism).

- The millions of different species of plants, animals, and microorganisms that live on earth today are related by descent from common ancestors.
- Biological classifications are based on how organisms are related. Organisms are classified into a hierarchy of groups and subgroups based on similarities that reflect their evolutionary relationships. The species is the most fundamental unit of classification.

Science Background for Teachers

One of the most common misconceptions about evolution is seen in the statement that "humans came from apes." This statement assumes that organisms evolve through a step-by-step progression from "lower" forms to "higher" forms of life and the direct transformation of one living species into another. Evolution, however, is not a progres-sive ladder. Furthermore, modern species are derived from, but are not the same as, organisms that lived in the past.

This activity has extensive historical roots. Few question the idea that Charles Darwin's *Origin of Species* in 1859 produced a scientific revolution. In essence, Darwin proposed a constellation of ideas that included: organisms of different kinds descended from a common ancestor (common descent); species multiply over time (speciation); evolution occurs through gradual changes in a population (gradualism); and competition among species for limited resources leads to differential survival and reproduction (natural selection). This activity centers on the theory of common descent. The theory of common descent was revolutionary because it introduced the concept of gradual evolution based on natural mechanisms. The theory of common descent also replaced a model of straight-line evolution with that of a branching model based on a single origin of life and subsequent series of changes—branching—into different species.

Based on his observations during the voyage of the H.M.S. *Beagle,* Darwin concluded that three species of mockingbirds on the Galapagos Islands must have some connection to the single species of mockingbird on the South American mainland. Here is the intellectual connection between observations and explanation. A species could produce multiple descendent species. Once this idea was realized, it was but a series of logical steps to the inferences that all birds, all vertebrates, and so on, had common ancestors.

Common descent has become a conceptual backbone for evolutionary biology. In large measure, this is so because common descent has significant explanatory power. Immediately, the idea found supporting evidence in comparative anatomy, comparative embryology, systematics, and biogeography. Recently, molecular biology has provided further support, as the students will discover in this activity. See Chapter 3 of this document and page 185 of the *National Science Education Standards* for more discussion of this topic.

This activity also introduces students to scientific evidence, models, and explanations as described in the accompanying excerpt drawn from the National Science Education Standards.

Evidence, Models, and Explanation

> *Evidence consists of observations and data on which to base scientific explanations. Using evidence to understand interactions allows individuals to predict changes in natural and designed systems.*
>
> *Models are tentative schemes or structures that correspond to real objects, events, or classes of events, and that have explanatory power. Models help scientists and engineers understand how things work. Models take many forms, including physical objects, plans, mental constructs, mathematical equations, and computer simulations.*
>
> *Scientific explanations incorporate existing scientific knowledge and new evidence from observations, experiments, or models into internally consistent, logical statements. Different terms, such as "hypothesis," "model," "law," "principle," "theory," and "paradigm," are used to describe various types of scientific explanations. As students develop and as they understand more science concepts and processes, their explanations should become more sophisticated. That is, their scientific explanations should*

more frequently include a rich scientific knowledge base, evidence of logic, higher levels of analysis, greater tolerance of criticism and uncertainty, and a clearer demonstration of the relationship between logic, evidence, and current knowledge

Materials and Equipment

For each student:

- Notebook
- Pencil

For each group of four students

- 4 sets of black, white, green, and red paper clips, each set with 35 paper clips

For the entire class:

- Overhead transparencies of *Characteristics of Apes and Humans,* Table 1, and *Morphological Tree,* Figure 1
- Overhead projector

Instructional Strategy: Part I

Engage Ask the students: When you hear the word "evolution," what do you think of first? Have the students explain what they understand about evolution. For many people, the first thing that comes to mind is often the statement "Humans evolved from apes."

Did humans evolve from modern apes, or do modern apes and humans have a common ancestor? Do you understand the difference between these two questions? This activity will give you the opportunity to observe differences and similarities in the characteristics of humans and apes. The apes discussed in this activity are the chimpanzee and the gorilla.

Explore Review Table 1, *Characteristics of Apes and Humans,* with the class. Make sure the students know that gibbons, chimpanzees, gorillas, and orangutans are four groups included in the ape family. Chimpanzees and gorillas represent the African side of the family; gibbons and orangutans represent the Asian side of the family. We focus only on the chimpanzee and gorilla in this activity. The only modern representative of the human family is *Homo sapiens,* although paleontologists have found fossil remains of other members, such as *Australopithecus afarensis* ("Lucy") and *Homo sapiens neandertalensis.*

Next discuss how the students can use the data to determine the relationships between humans, apes, and other animals. It might not be obvious that closely related organisms share more similarities than do distantly related organisms. Guide the students to the idea that structures might be similar because they carry out the same functions or because they were inherited from a common ancestor. Only those similarities that arise from a common ancestor can be used to determine evolutionary relationships.

Use the transparency of the *Morphological Tree,* Figure 1, for this discussion. Diagrams called branching trees illustrate relationships among organisms. One type of branching tree, called a morphological tree, is based on comparisons of skulls, jaws, skeletons, and other structures. Look carefully at the morphological tree.

Explain Ask the students to find the part of the morphological tree that shows the relationships between gorillas, chimpanzees, and humans. They will notice that there are no lines showing relationships. They should work with partners and develop three hypotheses to explain how these organisms are related. On a sheet of notebook paper, they should make a diagram of their hypotheses by drawing lines from Point A to each of the three organisms (G = gorilla, C = chimpanzee, H = human, A = common ancestor). Allow the students to develop their own hypotheses. Give them help only if you see they are not making any progress. Three hypotheses the students might propose are shown below (although not necessarily in the same order).

Instructional Strategy: Part II

Elaborate Modern research techniques allow biologists to compare the DNA that codes for certain proteins and to make predictions about the relatedness of the organisms from which they took the DNA. Students will use models of these techniques to test their hypotheses and determine which one is best supported by the data they develop.

Procedure Step 1. Working in groups of four, "synthesize" strands of DNA according to the following specifications. Each different color of paper clip represents one of the four bases of

DNA:

black = adenine (A)

green = guanine (G)

white = thymine (T)

red = cytosine (C)

Students should synthesize DNA strands by connecting paper clips in the proper sequence according to specifications listed for each group member. When they have completed the synthe-sis, attach a label to Position 1 and lay your strands on the table with Position 1 on the left. Each student will synthesize one strand of DNA. Thirty-five paper clips of each color should provide an ample assortment. To save time, make sure all strands are synthesized simultaneously. Emphasize to the students that they are using models to test the hypotheses they developed in the first part of the investigation. Following are directions for the respective groups:

- Group member 1

Synthesize a strand of DNA that has the following sequence:

Position 1 Position 20

A-G-G-C-A-T-A-A-A-C-C-A-A-C-C-G-A-T-T-A

Label this strand "human DNA." This strand represents a small section of the gene that codes for human hemoglobin protein.

- Group member 2

Synthesize a strand of DNA that has the following sequence:

Position 1 Position 20

A-G-G-C-C-C-C-T-T-C-C-A-A-C-C-G-A-T-T-A

Label this strand "chimpanzee DNA." This strand represents a small section of the gene that codes for chimpanzee hemoglobin protein.

- Group member 3

Synthesize a strand of DNA that has the following sequence:

Position 1 Position 20

A-G-G-C-C-C-C-T-T-C-C-A-A-C-C-A-G-G-C-C

Label this strand "gorilla DNA." This strand represents a small section of the gene that codes for gorilla hemoglobin protein.

- Group member 4

Synthesize a strand of DNA that has the following sequence:

Position 1 Position 20

A-G-G-C-C-G-G-C-T-C-C-A-A-C-C-A-G-G-C-C

Label this strand "common ancestor DNA." This DNA strand represents a small section of the gene that codes for the hemoglobin protein of a common ancestor of the gorilla, chimpanzee, and human. (You will use this strand in Part III.)

Emphasize to students that they will be using a model constructed from *hypothetical* data in the case of the common ancestor, since no such DNA yet exists, but that the other three sequences are real.

Step 2. Students should compare the human DNA to the chimpanzee DNA by matching the strands base by base (paper clip by paper clip).

Step 3. Students should count the number of bases that are not the same. Record the data in a table. Repeat these steps with the human DNA and the gorilla DNA.

The data for the hybridizations are as follows:

chimpanzee DNA, 5 unmatched bases; gorilla DNA, 10 unmatched bases. Be sure to ask the students to save all of their DNA strands for Part III.

Evaluate 1. How do the gorilla DNA and the chimpanzee DNA compare with the human DNA? The human DNA is more similar to the chimpanzee DNA than the gorilla DNA.

2. What do these data suggest about the relationship between humans, gorillas, and chimpanzees? The data suggest that humans are more closely related to the chimpanzee than they are to the gorilla.

3. Do the data support any of your hypotheses? Why or why not?

The data lend support to the hypothesis that the chimpanzee is more closely related to humans than the gorilla is.

4. What kinds of data might provide additional support for your hypotheses?

The students could test the hypotheses using additional data from DNA sequences or morphological features. They also could gather data from the fossil record.

Instructional Strategy: Part III

Begin this part by pointing out that biologists have determined that some mutations in DNA occur at a regular rate. They can use this rate as a "molecular clock" to predict when two organisms began to separate from a common ancestor. Most evolutionary biologists agree that humans, gorillas, and chimpanzees shared a common ancestor at one point in their evolutionary history. They disagree, however, on the specific relationships among these three species.

In this part of the activity, you will use data from your paper-clip model to evaluate different hypotheses about the relationships between humans, gorillas, and chimpanzees. Evolutionary biologists often disagree about the tempo of evolutionary change and about the exact nature of speciation and divergence. Reinforce the idea that models can be useful tools for testing hypotheses. Procedure Step 1. Assume that the common ancestor DNA synthesized in Part II represents a section of the hemoglobin gene of a hypothetical common ancestor. Compare this common ancestor DNA to all three samples of DNA (gorilla, human, and chimpanzee), one sample at a time. Record the data in a table.

The data for the comparisons are as follows: human DNA, 10 unmatched bases; chimpanzee DNA, 8 unmatched bases; gorilla DNA, 3 unmatched bases.

Evaluate 1. Which DNA is most similar to the common-ancestor DNA?

Gorilla DNA is most similar to the commonancestor DNA.

2. Which two DNAs were most similar in the way that they compared to the common-ancestor DNA? Human DNA and chimpanzee DNA have similar patterns when compared to the common ancestor DNA.

3. Which of the hypotheses developed in Part I do your data best support?

Answers will vary.

4. Do your findings prove that this hypothesis is correct? Why or why not?

Data from the models do not *prove* the validity of a hypothesis, but they do provide some direction for additional research.

5. Based on the hypothesis that your data best supported, which of the following statements is most accurate? Explain your answer in a short paragraph.

(a) Humans and apes have a common ancestor.

(b) Humans evolved from apes.

The students should infer that humans and apes share a common ancestor, represented by a com-mon branching point.

6. According to all the data collected, which of the following statements is most accurate? Explain your answer in a short paragraph.

(a) Chimpanzees and humans have a common ancestor.

(b) Chimpanzees are the direct ancestors of humans.

The students should infer that chimpanzees and humans share a common ancestor and that *modern* chimpanzees are not the direct *ancestors* of humans.

7. A comparison of many more DNA sequences indicates that human DNA and chimpanzee DNA are 98.8 percent identical. What parts of your data support this result?

The morphological tree and the DNA comparison data indicate that humans are closely related to chimpanzees.

8. What methods of science did you use in this activity?

Many answers are possible, including making observations, forming and testing hypotheses, and modeling.

ACTIVITY 5: PROPOSING EXPLANATIONS FOR FOSSIL FOOTPRINTS

In this activity, students observe and interpret "fossil footprint" evidence. From the evidence, they are asked to construct defensible hypotheses or explanations for events that took place in the geological past. The estimated time requirement for this activity is two class periods. This activity is designed for grades 5 through 8. The activity is adapted with permission from the Earth Science Curriculum Project.11

Standards-Based Outcomes

This activity provides all students an opportunity to develop the abilities of scientific inquiry and understanding of the nature of science as described in the *National Science Education Standards.* Specifically, it enables them to:

- propose explanations and make predictions based on evidence,
- recognize and analyze alternative explanations and predictions,
- understand that scientific explanations are subject to change as new evidence becomes available,
- understand that scientific explanations must meet certain criteria. First and foremost, they must be consistent with

experimental and observational evidence about nature, and must make accurate predictions, when appropriate, about systems being studied. They should also be logical, respect the rules of evidence, be open to criticism, report methods and procedures, and make knowledge public. Explanations of how the natural world changes based on myths, personal beliefs, religious values, mystical inspiration, superstition, or authority may be personally useful and socially relevant, but they are not scientific.

Science Background for Teachers

This activity provides teachers with the opportunity to help students realize the differences between observations and inferences. In terms of the *Standards,* it centers on the development of explanations based on evidence. It encourages students to think critically about the inferences they make and about the logical relationships between cause and effect.

Observations or statements of observations should have agreement by all individuals: "These are fossil footprints," or "The dimensions of one of the footprints is 20 cm by 50 cm." Inferences are statements that propose possible explanations for observations: "The two sets of footprints represent a fight between the animals." If this is true, then what evidence could you look for to support the inference. Note that the primary emphasis for this activity is developing abilities and understandings for "Science as Inquiry" as described in the *Standards.*12

Materials and Equipment

Make an overhead transparency of the footprint puzzle from the master provided on page 89. Have a blank piece of paper on hand to mask the puzzle when it is put on the projector.

Instructional Strategy

Engage Project position 1 of the footprints from the overhead by covering the other two positions with a blank piece of paper. Tell the students that tracks like these are common in parts of New England and in the southwestern United States. Point out to the students that they will be attempting to reconstruct happenings from the geological past by analyzing a set of fossilized tracks. Their problem is similar to that of a detective. They are to form defensible explanations of past events from limited evidence. As more evidence becomes available, their hypotheses must be modified or abandoned. The only clues are the footprints themselves. Ask the students: Can you tell anything about the size or nature of the organisms? Were all the tracks made at the same time? How many animals were involved? Can you reconstruct a series of events represented by this set of fossil tracks? Have the students discuss each of the questions. Accept any reasonable explanations students offer. Try consistently to point out the difference between what they observe and what they infer. Ask them to suggest evidence that would support their proposed explanations.

Explore Reveal the second position of the puzzle and allow time for the students to consider the new information. Students will see that the first explanation may need to be modified and new ones added.

Next project the complete puzzle and ask students to interpret what happened. A key point for students to recognize is that any reasonable explanation must be based only on those proposed explanations that still apply when all of the puzzle is projected. Any interpretation that is consistent with all the evidence is acceptable. Should it become necessary to challenge the students' thinking and stimulate the discussion, the following questions may help. Students should give evidence or suggest what they would look for as evidence to support their proposed explanations.

- In what directions did the animals move?
- Did they change their speed and direction?
- What might have changed the footprint pattern?
- Was the land level or irregular?
- Was the soil moist or dry on the day these tracks were made?
- In what kind of rock were the prints made?

- Were the sediments coarse or fine where the tracks were made?

The environment of the track area also should be discussed. If dinosaurs made the tracks, the climate probably was warm and humid. If students propose that some sort of obstruction prevented the animals from seeing each other, this might suggest vegetation. Or perhaps the widened pace might suggest a slope. Speculate on the condition of the surface at the time the footprints were made. What conditions were necessary for their preservation?

Explain An imaginative student should be able to propose several possible explanations. One of the most common is that two animals met and fought. No real reason exists to assume that one animal attacked and ate the other. Ask students who pro-pose this explanation to indicate the evidence. If they could visit the site, what evidence would they look for that would support their explanation. Certain lines of evidence—the quickened gaits, circular pattern, and disappearance of one set of tracks—could support the fight explanation. They might, however, support an explanation of a mother picking up her baby. The description and tempera-ment of the animals involved are open to question. Indeed, we lack the evidence to say that the tracks were made at the same time. The intermingling shown in the middle section of the puzzle may be evidence that both tracks were made at one time, but it could be only a coincidence. Perhaps one animal passed by and left, and then the other arrived. Discuss the expected learning outcomes related to scientific inquiry and the nature of science. To answer the questions posed by the set of fossil footprints, the students, like scientists, constructed reasonable explanations based solely on their logical interpretation of the available evidence. They recognized and analyzed alternative explanations by weighing the evidence and examining the logic to decide which explanations seemed most reasonable. Although there may have been several plausible explanations, they did not all have equal weight. In a manner similar to the way scientists work, students should be able to use scientific criteria to find, communicate, and defend the preferred explanation.

Elaborate You can have more discussions on interpreting series of events using animal prints students find outdoors and reproduce for the class. Do not forget to look for human footprints. Have students design a different fossil footprint puzzle. Choose several different ones and have student teams repeat the activity using the same learning goals.

Evaluate Describe a specific event involving two or more people or animals where footprint evidence remains. Ask the students, either in teams or individually, to diagram footprint evidence that could lead to several different, yet defensible, explanations regarding what took place. They should be able to explain the strengths and weaknesses of each explanation using their footprint puzzle.

ACTIVITY 6: UNDERSTANDING EARTH'S CHANGES OVER TIME

Comparing the magnitude of geologic time with spans of time in a person's lifetime is difficult for many students. In this activity, students use a long paper strip and a reasonable scale to represent visually all of geologic time, including significant events in the development of life on earth as well as recent human events. The investigation requires two class periods and is appropriate for grades 5 through 12. The activity is adapted with permission from the Earth Science Curriculum Project.13

Standards-Based Outcomes

This activity provides all students with an opportunity to develop understandings of the earth systems as described in the *National Science Education Standards.* Specifically, it introduces them to the following concepts:

- A mathematical scale representing the length of geologic time.
- The relationship of time between human events, events in earth's history, and the total age of the earth.
- The formation of the sun, the earth, and the rest of the solar system from a nebular cloud of dust and gas 4.6 billion years ago.

- The estimation of geologic time by observing rock sequences and using fossils to correlate the sequences at various locations. A current method of dating earth materials uses the known decay rates of radioactive isotopes present in rocks to measure the time since the rock was formed.
- The ongoing evolution of the earth system resulting from interactions among the solid earth, the oceans, the atmosphere, and organisms.
- The evidence for one-celled forms of life—the bacteria—extending back more than 3.5 billion years. The evolution of life caused dramatic changes in the composition of the earth's atmosphere, which did not originally contain oxygen.

Science Background for Teachers

Geologic time is largely subdivided on the basis of the evolution of life and on the amount and type of crustal activity that occurred in the past. Geologic time is ordered both relatively and absolutely. For relative dating the sequence in which rock strata formed is important; to explain the complete time scale for all of geologic history required correlating rock formations throughout the world. Fossils are important guides in this correlation as scientists assigned relative dates to the world's rocks according to a proposed sequence of life (fossil evidence).

Radiometric dating provides absolute ages for events in the earth's history. Radiometric dating techniques apply the decay rates of selected naturally radioactive isotopes to stable daughter isotopes to determine how long the unstable parent isotopes have been decaying. Fairly accurate dates have been determined for the events beginning in the Cambrian era; this comprises about 12 percent of the earth's history.

Materials and Equipment

The following materials will be needed by each group of two students:

- A paper strip, such as adding machine tape or shelf paper,

- A meter stick,
- Masking or cellophane tape.

If students use the scale suggested—1 millimeter to 1 million years and 1 meter to 1 billion years—a paper strip 5 meters in length efficiently accommodates the 4.5-billion-year time scale. Make copies of Student Investigation Sheet A on page 91 ("Approximate Ages of Events in Years Before the Present"). Students will use this page to conduct their investigation.

Instructional Strategy

Engage Ask students how long a million years is. How would students count or measure a million of anything? Use this discussion to help students arrive at the question: How does a million years, or even the time since the last ice age, compare with the age of the earth? Suppose you want to make a visual model showing a time line of the earth's history, how would you proceed?

Explore Provide students with Student Investigation Sheet A. Have them decide how to represent these events in a time-ordered sequence. Provide a roll of paper tape on which to plot the model. Students might need help in understanding how to set up a scale that can be displayed in the classroom or adjacent hallway. A reasonable scale is 1 millimeter to 1 million years, 1 centimeter to 10 million years, and 1 meter to 1 billion years. Depending on available space, larger unit distances will be easier to work with. Regardless of the scale the students choose, the last million years will be difficult to plot. Allow students to work out a scale on their own. However, to avoid undue confusion and frustration for some, review student progress after the first few minutes and be ready to ask leading questions or make suggestions. Allow students time to agree on a reasonable scale, mark the locations of each event on their time scale, and resolve the problem of trying to fit the events from the last 1 million years in the allotted space. When appropriate, encourage students to construct a separate, and larger, scale for mark-ing the most recent events.

Explain Discuss the long period of time in the earth's history before evidence of simple life forms, such as algae, appear in the fossil record. Note that time spans between significant "firsts" become shorter and shorter as you move closer and closer to "today." Compare and discuss expanded scales used to show more detail in the recent past. Discuss the role of scale in helping visualize and better understand the extremely long time span of the geologic time scale and the connections to biological evolution.

Elaborate Challenge students to develop an extended time scale to mark special events in their own lifetime and that of their parents, grandparents, or another adult. Have them calculate the percentage of the earth's history for which there is evidence of life, the percentage of the earth's histo-ry for which there is fossil evidence of the first humanlike animals, or the percentage of the earth's history during which dinosaurs lived.

Evaluate Ask students to calculate the length of a paper strip necessary to represent all of geologic time when using the extended scale they used to show the most recent events. Have students write a short news article explaining their scale of geologic time and the evolutionary changes in the earth's lithosphere, atmosphere, and biosphere.

ACTIVITY 7: PROPOSING THE THEORY OF BIOLOGICAL EVOLUTION: HISTORICAL PERSPECTIVE

This activity uses evolution to introduce students to historical perspectives and the nature of science. The teacher has students read short excerpts of original statements on evolution from Jean Lamarck, Charles Darwin, and Alfred Russel Wallace. This activity is intended as either a supplement to other investigations or as a core activity. Designed for grades 9 through 12, the activity requires a total of three class periods.

Standards-Based Outcomes

The activity provides all students with opportunities to develop understandings of the history and nature of science as described in the *National Science Education Standards.* Specifically, it con-veys the following concepts:

- Scientists are influenced by societal, cultural, and personal beliefs and ways of viewing the world. Science is not separate from society but rather a part of society.
- Scientific explanations must meet certain criteria.

First and foremost, they must be consistent with experimental and observational evidence about nature, and they must make accurate predictions, when appropriate, about the systems being studied.

They should also be logical, respect the rules of evidence, be open to criticism, report methods and procedures, and make knowledge public. Explanations of how the natural world changes based on myths, personal beliefs, religious values, mystical inspiration, superstition, or authority may be personally useful and socially relevant, but they are not scientific.

- Because all scientific ideas depend on experimental and observational confirmation, all scientific knowledge is in principle subject to change as new evidence becomes available. The core ideas of science, such as the conservation of energy or the laws of motion, have been subjected to a wide variety of confirmations and are therefore unlikely to change in the areas in which they have been tested. In areas where data or understanding are incomplete, such as the details of human evolution or questions surrounding global warming, new data may well lead to changes in current ideas or resolve current conflicts. In situations where information is still fragmentary, it is normal for scientif-ic ideas to be incomplete, but this is also where the opportunity for making advances may be greatest.
- Occasionally, there are advances in science and technology that have important and long-lasting effects on science and society.
- The historical perspective of scientific explanations demonstrates how scientific knowledge changes by evolving over time, almost always building on earlier knowledge.

Science Background for Teachers

In historical perspective, explanations for the origin and diversity of life are not new and probably began when humans first began asking questions about the natural world. By the time of the Greeks, individuals such as Thales (624 to 548 B.C.) and Anaximander (611 to 547 B.C.) proposed explanations for life's origins and gradual changes.

In the 1800s three individuals proposed explanations for biological evolution—Jean Lamarck, Charles Darwin, and Alfred Russel Wallace. In the early years of the nineteenth century, a French biol-ogist, Jean Lamarck (1744 to 1829), proposed a view of evolution that questioned the then popular idea that species did not change. Lamarck proposed the idea that changes do take place in animals over long periods of time, specifically through the use of organs and appendages. The popular example of Lamarck's idea is the long necks of giraffes that helped them feed higher in trees. Based on the extension and use of the neck, one generation of giraffes passed the longer neck to the next generation. (See the excerpt for this activity.) Charles Darwin (1809 to 1882) was born in England and completed his formal education at Cambridge University. Darwin's main interests centered on the study of nature and collecting a diversity of organisms. After graduation, Darwin's professor recommended him for the position of naturalist on H.M.S. *Beagle*. The voyage of the *Beagle* lasted five years (1831 to 1836) and provided the observations and evidence (in the form of specimens) that became the foundation for Darwin's theories. Of particular note in history is Darwin's observations on the Galapagos Islands located off the coast of Ecuador. Darwin's curiosity and insight led him to observe both similarities and differences among organisms and compare them on the mainland and the islands 600 miles offshore.

Based on his observations, he wondered about the origin of different plants and animals, and the variations in species he recorded in similar organisms.

After returning to England, Darwin spent more than twenty years studying the specimens, experimenting, and reviewing the notes of his voyage. In 1858 he was surprised to find that Alfred Russel Wallace had formulated similar conclusions. In the same year, Darwin reported his and Wallace's work in a joint presentation to the Scientific Society in London. One year later, in 1859, Darwin published On the Origin of Species by Means of Natural *Selection*. This publication caused great debate and what is now viewed as a scientific revolution. Darwin's theories of evolution have also had consid-erable impact on society and our cultural views. Alfred Russel Wallace (1823 to 1913) was also born in England. He became a teacher of English. He later developed an interest in collecting plants and insects. In 1848 he made an expedition to the Amazon River in Brazil to collect scientific materials. On a later expedition to the Malay Islands, Wallace observed some variations in organisms that engaged the same questions that Darwin posed— why did each island have different species? Wallace thought about the question for three years and in 1858 he proposed his theory.

Materials and Equipment

Excerpt from *Zoological Philosophy* by Jean Lamarck (provided)

Excerpt from *On the Tendency of Varieties to Depart Indefinitely from the Original Type* by Alfred Russel Wallace (provided)

Excerpt from *On the Origin of Species* by Charles Darwin (provided)

Instructional Strategy

These excerpts give the students an opportunity to read original statements by individuals who contributed to a major revolution in the history of biology. The instructional strategy is that of small-group discussions. Students read an original excerpt prior to class and discuss the reading in class.

Engage Introduce the sequence of readings by asking questions based on the learning outcomes:

- How do you think the society in which scientists live might influence their views?
- What makes a person's explanation scientific?

- Can scientific explanations change? If so, how? Why? If not, why not?
- Can you name some major theories in science?

In biology?

Ask the students what they know about the theory of evolution. What do they know about Charles Darwin? When did he propose his theory? Did any other individuals propose theories about evolution? How did Darwin develop his theory of evolution? Questions such as these will set the stage for the first reading. Assign the reading by Jean Lamarck as homework.

Explore Students should work in groups of four to discuss Jean Lamarck's explanations of changes in organisms. Questions for student discussions include:

- What is the role of the environment in Lamarck's explanation?
- What scientific approach is suggested by Lamarck's statement: "Nothing of all this can be considered as hypothesis or private opinion; on the contrary, they are truths which, in order to be made clear, only require attention and the observation of facts."
- Was Lamarck's explanation scientific? Why or why not?
- Can you propose any other explanations for Lamarck's observations about the disuse and use of organs?

Explain Prior to this group discussion, assign the reading by Alfred Russel Wallace. With your guidance, this discussion should clarify for students some of the fundamental concepts about science as a human endeavor and the nature of science. This should include discussion in groups of four followed by a full class summary of the learning outcomes.

- How would you characterize Wallace's idea that "The life of wild animals is a struggle for existence?" How is Wallace's view scientific?
- Wallace claims that "useful variations will tend to increase, unuseful or hurtful variations to diminish." How does this occur? What evidence does he cite?

- How does Wallace's explanation differ from Lamarck's?
- What do you think of Wallace's critique of Lamarck's hypotheses?

Elaborate Prior to this group discussion, assign the reading by Charles Darwin. In these discussions, students should apply concepts about the nature of science and the historical perspective developed during prior discussions. This discussion should demonstrate greater sophistication and understanding by the students.

- What led Darwin to formulate his ideas about the origin of species?
- On what did he base his explanations?
- What did Darwin propose as the origin of species?
- What was the relationship of Lamarck's and Wallace's work to Darwin's?
- Was Darwin's explanation scientific? Why or why not?
- How did Darwin attempt to determine how modifications of a species are accomplished?
- How did Darwin explain the incomplete nature of his ideas?

Evaluate Have each student write a brief essay on the nature of scientific knowledge as demonstrated in the development of the theory of evolution. They should cite at least two quotes from the reading to support their discussion. The essays should incorporate the concepts of adaptation, natural selection, and descent from common ancestors.

ACTIVITY 8: CONNECTING POPULATION GROWTH AND BIOLOGICAL EVOLUTION

In this activity, students develop a model of the mathematical nature of population growth.

The investigation provides an excellent opportunity for consideration of the population growth of plant and animal species and the resultant stresses that contribute to natural selection.

This activity will require two class periods and is appropriate for grades 5 through 12. The activity is based on an original activity from the Earth Science Curriculum Project. It is used with permission.

Standards-Based Outcomes

This activity provides all students an opportunity to develop understandings about scientific inquiry and biological evolution as described in the *National Science Education Standards.* Specifically, it conveys the following concepts:

- Mathematics is essential in scientific inquiry.

  Mathematical tools and models guide and improve the posing of questions, gathering data, constructing explanations, and communicating results.

- Species evolve over time. Evolution is the consequence of (1) the potential for a species to increase its numbers, (2) the genetic variability of offspring due to mutation and recombination of genes, (3) a finite supply of the resources required for life, and (4) the ensuing selection of those offspring better able to survive and leave offspring in a particular environment. (Item 1 is the primary content emphasis of this activity. Teachers can introduce the other factors as appropriate.)

- Populations grow or decline through the combined effects of births and deaths and through emigration and immigration into specific areas.

  Populations can increase through linear or exponential growth, with effects on resource use and on environmental pollution.

- Populations can reach limits to growth.

  Carrying capacity is the maximum number of organisms that can be supported by a given environment.

- Living organisms have the capacity to produce populations of arbitrarily large size, but environments and resources are finite. This fundamental tension has profound effects on the interactions between organisms.

Science Background for Teachers

The tension between expanding populations and limited resources was a fundamental point that Darwin came to understand when he read Thomas Malthus.15 This understanding subsequently had an important influence on the formulation of his theory of natural selection.

This activity extends the general idea of population growth to humans. Here the important point is that human beings live within the world's ecosystems. Increasingly, humans modify ecosys-tems as a result of population growth, technology, and consumption.

Human destruction of habitats through direct harvesting, pollution, atmospheric changes, and other factors is threatening current global stability, and, if not addressed, ecosystems will be irreversibly affected.

The increase in the size of a population (such as the human population) is an example of exponential growth. The human population grew at the slow rate of only about 0.002 percent a year for the first several million years of our existence. Since then the average annual rate of human population has increased to an all-time high of 2.06 percent in 1970. As the base number of people undergoing growth has increased, it has taken less and less time to add each new billion people. It took 2 million years to add the first billion people; 130 years to add the second billion; 30 years to add the third billion; 15 years to add the fourth billion; and only 12 years to add the fifth billion. We are now approaching the sixth billion.

MATERIALS AND EQUIPMENT

Each group of three or four students will need:

- Approximately 2,000 small, uniformly shaped objects (kernels of corn, dried beans, wooden markers, plastic beads)
- 10 paper cups or small beakers
- A 250-ml or 400-ml beaker

Instructional Strategy

Engage Initiate a discussion on human population with such questions as: How long have humans been on the earth? How do you think the early rate of human population growth compares with the population growth rate today? Why did this rate change? Tell students that this investigation represents a model of population growth rates.

Explore Have student groups complete the following activities.

- Place the glass beakers on their desks. Begin by placing two objects (e.g., corn or plastic beads) in it. The beaker represents the limits of an ecosystem or ultimately the earth.
- Place 10 cups in a row on their desk. In the first cup, place two objects. In the second cup, place twice as many objects as the first cup (four). Have students record the number of objects on the outside of the cup. Continue this procedure by placing twice as many objects as in the former cup, or doubling the number, in cups 3 through 10. Be sure students record the numbers on the cups.
- Take the beaker and determine its height. Have students indicate the approximate percentage of volume that is *without* objects. Record this on the table as 0 time.
- At timed intervals of 30 seconds, add the contents of cups 1 through 10. Students should record the total population and approximate percentage of volume in the beaker that is without objects.
- Students should complete the procedure and graph their results as total population versus results.

Students may question the need for the 30-second intervals. The length of the time interval is arbitrary. Any time interval will do. Preparation of the graph can be assigned as homework.

Range of Results

The mathematics involved in answering the questions may challenge some students. Assist students when necessary to enable them to accomplish the objectives of the investigation. Table below shows the population and the percent of the beaker's volume without objects. A typical student graph is shown in Figure 1.

Explain Ask the students to explain the relationship between population growth and biological evolution in populations of microorganisms, plants, and animals. Through questions and discussion, help them develop the connections stated in the learning outcome for the activity. Evolution results from an interaction of factors related to the potential for species to increase in numbers, the genetic variability in a population, the supply of essential resources, and environmental pressures for selection of those offspring that are able to survive and reproduce.

Table: Population growth

Time Internal	Population	Percentage of empty volume (400-ml beaker)
0	2	99%
1	4	99%
2	8	99%
3	16	98%
4	32	97%
5	64	95%
6	128	93%
7	256	88%
8	512	80%
9	1024	70%
10	2048	50%
11	4096	0%

Elaborate Begin by having students explain the results of their activity. During the discussion of the graph, have the students consider some of the following: Are there any limitations to the number of people the earth will support? Which factor might limit population growth first? How does this factor relate to human evolution? Are there areas in the world where these limits have been reached already? Have we gone beyond the earth's ideal population yet? What problems will we face if we overpopulate the earth? How might human influence on, for example, habitats affect biological evolution. Students' answers to these questions will vary, depending on their background and information. The outcome, however, should be an intense discussion of some vital problems and should provide opportunities to introduce the fundamental concepts from the *National Science Education Standards.*

Evaluation 1. Human population on the earth is thought to have had a slow start, with doubling periods as long as 1 million years. The current world population is thought to be doubling every 37 years. How would this growth rate compare with the rates found in your investigation? Both the population in the investigation and on the earth increase in a geometric progression. This means the graphs have the same shape. You can substitute 37 years for every 30-second interval and the numbers will represent actual world population growth. The slope of the graph would remain the same.

2. What happens to populations when they reach the limits to growth? The populations stop growing because death rates (or emigration) exceed birth rates (or immigration).

SELECTING INSTRUCTIONAL MATERIALS

Quality instructional materials are essential in teaching about evolution and the nature of science. It also is important to consider the context within which specific materials will be used. This chapter therefore begins with brief discussions of school science programs and the criteria used to design curricula.

CRITERIA FOR CONTEMPORARY SCIENCE CURRICULUM

Before selecting specific materials to teach evolution and the nature of science, it is important to identify criteria that can help evaluate school science programs and the design of instructional materials. Chapter seven in the *National Science Education Standards,* "Science Education Program Standards," describes the conditions needed for quality school science programs. These conditions focus on six areas:

- Consistency across all elements of the science program and across the K-12 continuum
- Quality in the program of studies
- Coordination with mathematics
- Quality resources
- Equitable opportunities for achievement
- Collaboration within the school community to support a quality program

Similarly, educators need to consider criteria against which to judge instructional materials. Teachers, curriculum designers, and other school personnel can use the following criteria to evaluate the design of a new curriculum, to select instructional materials, or to adapt instructional materials through professional development. No set of instructional materials will meet all the following criteria. You will have to make a judg-ment about the degree to which materials meet criteria and about acceptable and unacceptable omissions. These criteria are adapted from earlier discussions of standards-based curriculum.1

Criterion 1: *A Coherent, Consistent, and* Coordinated Framework for Science Content. Science content should be consistent with national, state, and local standards and benchmarks. Whether for lessons, units, or a complete elementary, middle, or high school program, the content should be well-thought-out, coordinated, and conceptually, procedurally, and coherently organized. The roles of science concepts, inquiry, science in personal and social contexts, and the history and nature of science should be clear and explicit.

Criterion 2: *An Organized and Systematic Approach to Instruction.* Most contemporary science curricula incorporate an instructional model. The instructional model should (1) provide for different forms of interaction among students and between the teachers and students, (2) incorporate a variety of teaching strategies, such as inquiry-oriented investigations, cooperative groups, use of technology, and (3) allow adequate time and opportunities for students to acquire knowledge, skills, and attitudes.

Criterion 3: *An Integration of Psychological* Principles Relative to Cognition, Motivation, *Development, and Social Psychology.* Psychological principles such as those found in the American Psychological Association publication *How Students* Learn: Reforming School Through Learner-*Centered Education*2 should be applied to the framework for content, teaching, and assessment. These psychological principles include more than learning theory. They include providing for motiva-tion, development, and social interactions.

Criterion 4: *Varied Curriculum Emphases.* The idea of curriculum emphases can be expressed by thinking about the foreground and background in a painting. An artist decides what will be in the foreground, and that subject is emphasized. Science curricula can, for example, emphasize science concepts, inquiry, or the history and nature of science, while other goals may be evident but not emphasized. No one curriculum emphasis is best for all students; probably, a variety of emphases accom-modates the interests, strengths, and demands of science content.

Criterion 5: *An Array of Opportunities to* Develop Knowledge, Understanding, and Abilities Associated with Different Dimensions of Scientific *Literacy.* Contemporary science curricula should provide a balance among the different dimensions of science literacy, which include an understanding of scientific concepts, the ability to engage in inquiry, and a capacity to apply scientific informa-tion in making decisions.3

Criterion 6: *Teaching Methods and Assessment* Strategies Consistent with the Goal of Science *Literacy.* Approaches to teaching and assessment ought to be consistent with the goals of teaching evolution, inquiry, and the history and nature of science. This can be accomplished by using inquiry-oriented teaching methods and by assessing students during investigative activities.

Criterion 7: *Professional Development for* Science Teachers Who Implement the Curriculum. Curricula need to provide opportunities that support teachers as they develop the knowledge and skills associated with implementing and institutionalizing the science program.

Criterion 8: *An Inclusion of Appropriate Educational Technologies.* The use of computers and various types of software enhances learning when students use the technologies in meaningful ways. The use of educational technologies should be consistent with other features of the curriculum— for instance, the dimensions of scientific literacy and an instructional model.

Criterion 9: *Thorough Field Testing and Review* for Scientific Accuracy and Pedagogic Quality. One important legacy of the 1960s curriculum reform is the field testing of materials in a variety of science classrooms.

Field testing and reviewing a program identify problems that developers did not recognize and fine tune the materials to the varied needs of teachers, learners, and schools. Scientists should review materials for accuracy.

Developers can miss the subtleties of scientific concepts, inquiry, and design. In addition, educators who review materials can provide valuable insights about teaching and assessment that help developers improve materials and enhance learning.

Criterion 10: *Support from the Educational System.* Research on the adoption, implementation, and change associated with curricula indicates the importance of intellectual, financial, and moral support from those within the larger educational system.4 This support includes science teachers, administrators, school boards, and communities.

Although a curriculum cannot ensure support, it should address the need for support and provide indicators of support, such as provision of materials and equipment for laboratory investigations, budget allocations for professional development, and proclamations by the school board.

Clearly, no one curriculum thoroughly incorpo-rates all ten criteria. There are always trade-offs when developing, adapting, or adopting a science curriculum. However, the criteria should provide assistance to those who have the responsibility of improving the science curriculum.

ANALYZING INSTRUCTIONAL MATERIALS

The process of selecting quality materials includes determining the degree to which they are consistent with the goals, principles, and criteria developed in the *National Science Education Standards*. Well-defined selection criteria help ensure a thoughtful and effective process. To be both usable and defensible, the selection criteria must be few in number and embody the critical tenets of accurate science content, effective teaching strategies, and appropriate assessment techniques.

The process described in the following pages can help teachers, curriculum designers, or other school personnel complete a thorough and accurate evaluation of instructional materials. To help make this examination both thorough and usable, references to specific pages and sections in the *National Science Education Standards* have been provided, as have worksheets to keep track of the information needed to analyze and select the best instructional materials.

ANALYSIS PROCEDURES

The procedures outlined in this section include:

- Overview of instructional materials
- Analysis of science subject matter
- Analysis of pedagogy
- Analysis of assessment process
- Evaluating the teacher's guide
- Analysis of use and management

The extent to which instructional materials meet the criteria outlined in this chapter determines their usefulness for classroom teachers and the degree of alignment with the *Standards*. A thorough analysis of instructional materials requires considerable time and collaboration with others and attention to detail. Good working notes are helpful in this process. We recommend using the analysis worksheets provided at the end of this chapter.

Overview of Instructional Materials

The following overview of instructional materials introduces the review process and provides a general context for analysis and subsequent selection of specific materials.

1. The first consideration is whether the key concepts of evolution and the nature of science are being emphasized. To help make this determina-tion, locate the table of contents, index, and glossary in the material you are evaluating. The box below contains terms related to fundamental concepts in evolution and the nature of science taken from the *Standards*. Record page numbers where each is found for future reference. (See Worksheet 1 on page 112 in the back of this chapter.) These terms will give you a preliminary indication of coverage on these fundamental topics.

   Evolution: evolution, diversity, adaptation, interpreting fossil evidence, techniques for age determination, natural selection, descent from common ancestors

   Nature of Science: explanation, experiment, evidence, inquiry, model, theory, skepticism

2. Look through both student and teacher materials. Are student outcomes listed? Note page numbers for several outcomes related to evolution and the nature of science.
3. Look for student investigations or activities. Where are they located? Note that in some materials, student investigations are integrated within the reading material. In others they are located in a separate section—sometimes at the back of a chapter or book or in a separate laboratory manual.
4. Read several relevant paragraphs of student text material. What is your judgment about the concepts? Are the concepts in the students' text consistent with the fundamental concepts in the *Standards*? Does the text include more, fewer, or different concepts?
5. Do the photographs and illustrations provide further understanding of the fundamental concepts?

Analysis of Instructional Materials for Science Subject Matter

A. Content

The following procedures for content analysis will help you examine instructional materials for fundamental concepts of evolution, science as inquiry, and the nature of science. Look for evi-dence in discussions in the text and in the student investigations to determine the degree to which the fundamental concepts are addressed.

Fundamental concepts underlying specific standards on evolution and the nature of science are referenced below. (Note: You will need a copy of the *National Science Education Standards* or access to it through the World Wide Web at www.nap.edu/readingroom/books/nses.)

1. Choose a lesson or representative section of the student instructional materials on the topic of evolution. Make a preliminary list of the fundamental concepts from the *Standards* that are included in the lesson and place them on your worksheet. (See Worksheet 2 on page 114 in the back of this chapter.)
2. Select one of these fundamental concepts and list all sections of the materials that deal with this idea. Determine whether the materials focus on the fundamental concepts, or if they represent only a superficial match. For example, Life Science Standard C in the *Standards5* specifies: "Biological evolution accounts for the diversity of species developed through gradual processes over many generations. Species acquire many of their unique characteristics through biological adaptation, which involves the selection of naturally occurring variations in populations." The instructional materials should provide opportunities for students to develop an understanding of biodiversity and evolution as described in the *Standards*. A negative example would be defining the term biodiversity only in reference to the fact that wide varieties of plants and animals populate particular environments.

You should complete this analysis for all fundamental concepts associated with a particular standard. The more fundamental concepts you analyze using this process, the more confidence you will have in the quality of the instructional materials and their alignment with the *Standards*. Identify the fundamental concepts that are not developed and the variation of treatment among those that are included in the materials.

3. If appropriate, select one of the student investigations for analysis of subject matter. On what fundamental concepts from Life Science Standard C or Earth and Space Science Standard D is the investigation focused? To what degree does the activity fulfill the intent of the fundamental concepts? For example, making and comparing model casts and molds of sea shells does not necessarily contribute to an understanding of how fossils are formed or provide important evidence of how life and environmental conditions have changed. It is recommended that you analyze a second student investigation.

B. Scientific Inquiry

1. You should develop some understanding of scientific inquiry in the *Standards*. Read Standard A, Science as Inquiry, referenced on the following page.

   Note that Standard A specifies two separate aspects of science as inquiry: abilities necessary to do scientific inquiry, and fundamental understand-ings about scientific inquiry. Examine several lessons in the student and teacher materials to answer the following question: To what degree do the lessons provide students the opportunity to develop the abilities and understandingsof scien-tific inquiry?

2. Read through the text narrative, looking for student investigations and examining any suggestions for activities outside of class time. Are opportunities provided for students to develop abilities of scientific inquiry such as posing their own relevant questions, planning and conducting investigations, using appropriate tools and

techniques to gather data, using evidence to communicate defensible explanations of cause and effect relationships, or using scientific criteria to analyze alternative explanations to determine a preferred explanation? Record page numbers where examples are found and make notes of explanation.

2. What opportunities are provided for students to develop a fundamental understanding of scientific inquiry? In addition to the language of the text, examine the teacher's guide for suggestions that teachers can use to discuss the role and limitations of scientific skills such as making observations, organizing and interpreting data, and constructing defensible explanations based on evidence. Can you find a discussion of how science advances through legitimate skepticism? Can you find a discussion of how scientists evaluate proposed explanations of others by examining and comparing evidence, identifying reasoning that goes beyond the evidence, and suggesting alternative explanations for the same evidence? Are there opportunities for students to demonstrate these same understandings as a part of their investigations? Make notes where this evidence is found for later reference.

C. History and Nature of Science

1. Are history and the nature of science incorporated into the treatment of evolution? Read Standard G, History and Nature of Science, referenced in the following box.
2. Read through several lessons in the student and teacher materials. Can you find examples describing the roles of scientists, human insight, and scientific reasoning in the historical and contemporary development of explanations for evolution? Can you find specific references to historical contributions of scientists in the development of fundamental concepts of evolution? What evidence can you find in the text narrative or student investigations that demonstrates how scientific explanations are developed, reviewed by peers, and revised in light of new evidence and thinking?

Analysis of Pedagogy

What students learn about evolution and the nature of science depends on many things, including the accuracy and developmental appropriateness of content and its congruence with the full intent of the content standards. Opportunities to learn should be consistent with contemporary models of learning.

The criteria in this section are based on characteristics of effective teaching proposed in Teaching Standards A, B, and E. Using the following sequence of questions, examine several lessons in the student materials and the teacher's guide. (See Worksheet 3 on page 117 in the back of this chapter.)

1. Do the materials identify specific learning goals or outcomes for students that focus on one or more of the fundamental concepts of evolution and the nature of science?
2. Study the opening pages of a relevant chapter or section. Does the material on the opening pages of the chapter or section on evolution engage and focus student thinking on interesting questions, problems, or relevant issues?
3. Does the material provide a sequence of learning activities connected in such a way as to help students build understanding of a fundamental concept? Are suggestions provided to help the teacher keep students focused on the purpose of the lesson?
4. Does the teacher's guide present common stu-dent misconceptions related to the fundamental concepts of evolution and the nature of science? Are suggestions provided for teachers to find out what their students already know? Are there learning activities designed to help students confront their misconceptions and encourage conceptual change?

Analysis of Assessment Process

Assessment criteria in this section are grounded in the Assessment Standards. Examine several lessons in the student and teacher materials for evidence to answer the following questions. (See Worksheet 4 on page 118 in the back of this chapter.)

1. Is there consistency between learning goals and assessment? For example, if instruction focuses on building understanding of fundamental concepts, do assessments focus on explanations and not on vocabulary?
2. Do assessments stress application of concepts to new or different situations? For example, are the students asked to explain new situations with concepts they have learned?
3. Are assessment tasks fair for all students? For example, does success on assessment tasks depend too heavily on the student's ability to read complex items or write explanations as opposed to understanding the fundamental concepts?
4. Are suggestions for scoring criteria or rubrics provided for the teacher?

Evaluating the Teacher's Guide

Examine several lessons in the teacher's guide to help answer the following questions:

1. Does the teacher's guide present appropriate and sufficient background on science?
2. Are the suggested teaching strategies usable by most teachers?
3. Are suggestions provided for pre-and postinvestigation discussions focusing on concept development, inquiry, and the nature of science?
4. Does the teacher's guide recommend addi-tional professional development?
5. Does the teacher's guide indicate the types of support teachers will need for the instructional materials?

Analysis of Use and Management

A high degree of alignment with *Standards* content, pedagogy, and assessment criteria does not necessarily guarantee that instructional materials will be easy to manage. The *Standards* address the importance of professional development, and some aspects of the program standards apply as well.6

1. How many different types of materials must be managed and orchestrated during a typical chapter, unit, or teaching sequence (e.g., student text, teacher's guide, transparencies, handouts, videos, and software)? (See Worksheet 5 on page 119 in the back of this chapter.)
2. Does the teacher's guide contain suggestions for effectively managing materials?
3. Do the instructional materials call for equipment, supplies, and technology that teachers may not have?
4. Do the instructional materials identify safety issues and provide adequate precautions?
5. Is the cost for materials and replacements reasonable? Are there special requirements?

5

THE THEORY OF BIOLOGICAL SCIENCE

Science is a particular way of knowing about the world. In science, explanations are limited to those based on observations and experiments that can be substantiated by other scientists. Explanations that cannot be based on empirical evidence are not a part of science.

In the quest for understanding, science involves a great deal of careful observa-tion that eventually produces an elaborate written description of the natural world. Scientists communicate their findings and conclusions to other scientists through publications, talks at conferences, hallway conversations, and many other means. Other scientists then test those ideas and build on preexisting work. In this way, the accuracy and sophistication of descriptions of the natural world tend to increase with time, as subsequent generations of scientists correct and extend the work done by their predecessors.

Progress in science consists of the development of better explanations for the causes of natural phenomena. Scientists never can be sure that a given explanation is complete and final. Some of the hypotheses advanced by scientists turn out to be incorrect when tested by further observations or experiments. Yet many scientific explanations have been so thoroughly tested and confirmed that they are held with great confidence.

The theory of evolution is one of these well-established explanations. An enor-mous amount of scientific investigation since the mid-19th century has converted early ideas about evolution proposed by Darwin and others into a strong and well-supported theory. Today, evolution is an extremely active field of research, with an abundance of new discoveries that are continually increasing our understanding of how evolution occurs.

This booklet considers the science that supports the theory of evolution, focusing on three categories of scientific evidence:

- Evidence for the origins of the universe, Earth, and life
- Evidence for biological evolution, including findings from paleontology, comparative anatomy, biogeography, embryology, and molecular biology
- Evidence for human evolution

At the end of each of these sections, the positions held by advocates of "creation science" are briefly presented and analyzed as well. The theory of evolution has become the central unifying concept of biology and is a critical component of many related scientific disciplines. In contrast, the claims of creation science lack empirical support and cannot be meaningfully tested. These observations lead to two fundamental conclusions: the teaching of evolution should be an integral part of science instruction, and creation science is in fact not science and should not be presented as such in science classes.

TERMS USED IN DESCRIBING THE NATURE OF SCIENCE*

Fact: In science, an observation that has been repeatedly confirmed and for all practical purposes is accepted as "true." Truth in science, however, is never final, and what is accepted as a fact today may be modified or even discarded tomorrow.

Hypothesis: A tentative statement about the natural world leading to deductions that can be tested. If the deductions are verified, it becomes more probable that the hypothesis is correct. If the deductions are incorrect, the original hypoth-esis can be abandoned or modified. Hypotheses can be used to build more complex inferences and explanations.

Law: A descriptive generalization about how some aspect of the natural world behaves under stated circumstances.

Terms Used in Describing the Nature of Science*

Theory: In science, a well-substantiated explanation of some aspect of the natural world that can incorporate facts, laws, inferences, and tested hypotheses. The contention that evolution should be taught as a "theory, not as a fact" confuses the common use of these words with the scientific use. In science, theo-ries do not turn into facts through the accumulation of evidence. Rather, theories are the end points of sci-ence. They are understandings that develop from extensive observation, experimentation, and creative reflection. They incorporate a large body of scientific facts, laws, tested hypotheses, and logical inferences. In this sense, evolution is one of the strongest and most useful scientific theories we have.

*Adapted from *Teaching About Evolution and the Nature of Science* by the National Academy of Sciences (Washington, D.C.: National Academy Press, 1998).

THE ORIGIN OF THE UNIVERSE, EARTH, AND LIFE

The term "evolution" usually refers to the biological evolution of living things. But the processes by which planets, stars, galaxies, and the universe form and change over time are also types of "evolution." In all of these cases there is change over time, although the processes involved are quite different. In the late 1920s the American astronomer Edwin Hubble made a very interest-ing and important discovery. Hubble made observations that he interpreted as showing that distant stars and galaxies are receding from Earth in every direction.

Moreover, the velocities of recession increase in proportion with distance, a discov-ery that has been confirmed by numerous and repeated measurements since Hubble's time. The implication of these findings is that the universe is expanding. Hubble's hypothesis of an expanding universe leads to certain deductions. One is that the universe was more condensed at a previous time.

From this deduction came the suggestion that all the currently observed matter and energy in the uni-verse were initially condensed in a very small and infinitely hot mass. Ahuge explosion, known as the Big Bang, then sent matter and energy expanding in all directions. This Big Bang hypothesis led to more testable deductions. One such deduction was that the temperature in deep space today should be several degrees above absolute zero. Observations showed this deduction to be correct. In fact, the Cosmic Microwave Background Explorer (COBE) satellite launched in 1991 con-firmed that the background radiation field has exactly the spectrum predicted by a Big Bang origin for the universe.

As the universe expanded, according to current scientific understanding, matter collected into clouds that began to condense and rotate, forming the forerunners of galaxies. Within galaxies, including our own Milky Way galaxy, changes in pressure caused gas and dust to form distinct clouds. In some of these clouds, where there was sufficient mass and the right forces, gravitational attraction caused the cloud to collapse. If the mass of material in the cloud was sufficiently compressed, nuclear reactions began and a star was born.

Some proportion of stars, including our sun, formed in the middle of a flattened spinning disk of material. In the case of our sun, the gas and dust within this disk collided and aggregated into small grains, and the grains formed into larger bodies called planetesimals ("very small planets"), some of which reached diameters of several hun-dred kilometers. In successive stages these planetesimals coalesced into the nine planets and their numerous satel-lites.

The rocky planets, including Earth, were near the sun, and the gaseous planets were in more distant orbits. The ages of the universe, our galaxy, the solar system, and Earth can be estimated using modern scientific methods. The age of the universe can be derived from the observed relationship between the velocities of and the distances separating the galaxies. The velocities of distant galaxies can be measured very accurately, but the measurement of distances is more uncertain. Over the past few decades, measurements of the Hubble expansion have led to estimated ages for the universe of between 7 billion and 20 billion years, with the most recent and best measurements within the range of 10 billion to 15 billion years.

The age of the Milky Way galaxy has been calculated in two ways. One involves studying the observed stages of evolution of different-sized stars in globu-lar clusters. Globular clusters occur in a faint halo surrounding the center of the Galaxy, with each cluster containing from a hundred thousand to a million stars. The very low amounts of elements heavier than hydrogen and helium in these stars indicate that they must have formed early in the history of the Galaxy, before large amounts of heavy elements were created inside the initial generations of stars and later distributed into the interstellar medium through supernova explosions (the Big Bang itself created primarily hydrogen and helium atoms). Estimates of the ages of the stars in globular clusters fall within the range of 11 billion to 16 billion years. A second method for estimating the age of our galaxy is based on the present abundances of several long-lived radioactive elements in the solar system. Their abundances are set by their rates of production and distribution through exploding supernovas. According to these calculations, the age of our galaxy is between 9 bil-lion and 16 billion years. Thus, both ways of estimating the age of the Milky Way galaxy agree with each other, and they also are consistent with the independently derived estimate for the age of the universe.

Radioactive elements occurring naturally in rocks and minerals also provide a means of estimating the age of the solar system and Earth. Several of these elements decay with half lives between 700 million and more than 100 billion years (the half life of an element is the time it takes for half of the element to decay radioactively into another element). Using these time-keepers, it is calculated that meteorites, which are fragments of asteroids, formed between 4.53 billion and 4.58 billion years ago (asteroids are small "planetoids" that revolve around the sun and are remnants of the solar nebula that gave rise to the sun and planets). The same radioactive time-keepers applied to the three oldest lunar samples returned to Earth by the Apollo astronauts yield ages between 4.4 billion and 4.5 billion years, provid-ing minimum estimates for the time since the formation of the moon. The oldest known rocks on Earth occur in northwestern Canada (3.96 billion years), but well-studied rocks nearly as old are also found in other parts of the world. In Western Australia, zircon crystals encased within younger rocks have ages as old as 4.3 billion years, making these tiny crystals the oldest materials so far found on Earth.

The best estimates of Earth's age are obtained by calculating the time required for development of the observed lead isotopes in Earth's oldest lead ores. These estimates yield 4.54 billion years as the age of Earth and of meteorites, and hence of the solar system.

The origins of life cannot be dated as precisely, but there is evidence that bacteria-like organisms lived on Earth 3.5 billion years ago, and they may have existed even earlier, when the first solid crust formed, almost 4 billion years ago. These early organisms must have been simpler than the organisms living today. Furthermore, before the earliest organisms there must have been structures that one would not call "alive" but that are now components of living things. Today, all liv-ing organisms store and transmit hereditary information using two kinds of molecules: DNA and RNA. Each of these molecules is in turn composed of four kinds of subunits known as nucleotides. The sequences of nucleotides in particular lengths of DNA or RNA, known as genes, direct the construction of molecules known as proteins, which in turn catalyze biochemical reactions, provide structural compo-nents for organisms, and perform many of the other functions on which life depends. Proteins consist of chains of subunits known as amino acids. The sequence of nucleotides in DNA ~nd RNAtherefore determines the sequence of amino acids in proteins; this is a central mechanism in all of biology.

Experiments conducted under conditions intended to resemble those present on primitive Earth have resulted in the production of some of the chemical components of proteins, DNA, and RNA. Some of these molecules also have been detected in meteorites from outer space and in interstellar space by astronomers using radio-telescopes. Scientists have concluded that the "building blocks of life" could have been available early in Earth's history.

An important new research avenue has opened with the discovery that certain molecules made of RNA, called ribozymes, can act as catalysts in modern cells. It previously had been thought that only proteins could serve as the catalysts required to carry out specific biochemical functions. Thus, in the early prebiotic world, RNA molecules could have been "autocatalytic"—that is, they could have replicated themselves well before there were any protein catalysts (called enzymes).

Laboratory experiments demonstrate that replicating autocatalytic RNAmolecules undergo spontaneous changes and that the variants of RNAmolecules with the greatest autocatalytic activity come to prevail in their environments. Some scientists favor the hypothesis that there was an early "RNAworld," and they are testing models that lead from RNAto the synthesis of simple DNAand protein molecules.

These assemblages of molecules eventually could have become packaged within membranes, thus making up "protocells"—early versions of very simple cells. For those who are studying the origin of life, the question is no longer whether life could have originated by chemical processes involving nonbiological components. The question instead has become which of many pathways might have been followed to produce the first cells.

Will we ever be able to identify the path of chemical evolution that succeeded in initiating life on Earth? Scientists are designing experiments and speculating about how early Earth could have provided a hospitable site for the segregation of molecules in units that might have been the first living systems.

The recent specu-lation includes the possibility that the first living cells might have arisen on Mars, seeding Earth via the many meteorites that are known to travel from Mars to our planet.

Of course, even if a living cell were to be made in the laboratory, it would not prove that nature followed the same pathway billions of years ago. But it is the job of science to provide plausible natural explanations for natural phenomena. The study of the origin of life is a very active research area in which important progress is being made, although the consensus among scientists is that none of the current hypotheses has thus far been confirmed. The history of science shows that seemingly intractable problems like this one may become amenable to solution later, as a result of advances in theory, instrumentation, or the discovery of new facts.

Creationist Views of the Origin of the Universe, Earth, and Life

Many religious persons, including many scientists, hold that God created the universe and the various processes driving physical and biological evolution and that these processes then resulted in the creation of galaxies, our solar system, and life on Earth. This belief, which sometimes is termed "theistic evolution," is not in disagreement with scientific explanations of evolution. Indeed, it reflects the remarkable and inspiring character of the physical universe revealed by cosmology, paleontology, molecular biology, and many other scientific disciplines. The advocates of "creation science" hold a variety of viewpoints. Some claim that Earth and the universe are relatively young, perhaps only 6,000 to 10,000 years old. These individuals often believe that the present physical form of Earth can be explained by "catastrophism," including a worldwide flood, and that all living things (including humans) were created miraculously, essentially in the forms we now find them.

Other advocates of creation science are willing to accept that Earth, the planets, and the stars may have existed for millions of years. But they argue that the various types of organisms, and especially humans, could only have come about with supernatural intervention, because they show "intelligent design." In this booklet, both these "Young Earth" and "Old Earth" views are referred to as "creationism" or "special creation."

There are no valid scientific data or calculations to substantiate the belief that Earth was created just a few thousand years ago.

This document has summarized the vast amount of evidence for the great age of the universe, our galaxy, the solar system, and Earth from astronomy, astrophysics, nuclear physics, geology, geochem-istry, and geophysics. Independent scientific methods consistently give an age for Earth and the solar system of about 5 billion years, and an age for our galaxy and the universe that is two to three times greater. These conclusions make the origin of the universe as a whole intelligible, lend coherence to many different branches of science, and form the core conclusions of a remarkable body of knowledge about the origins and behavior of the physical world.

Nor is there any evidence that the entire geological record, with its orderly suc-cession of fossils, is the product of a single universal flood that occurred a few thou-sand years ago, lasted a little longer than a year, and covered the highest mountains to a depth of several meters. On the contrary, intertidal and terrestrial deposits demonstrate that at no recorded time in the past has the entire planet been under water. Moreover, a universal flood of sufficient magnitude to form the sedimentary rocks seen today, which together are many kilometers thick, would require a volume of water far greater than has ever existed on and in Earth, at least since the formation of the first known solid crust about 4 billion years ago. The belief that Earth's sediments, with their fossils, were deposited in an orderly sequence in a year's time defies all geological observations and physical principles concerning sedimentation rates and possible quantities of suspended solid matter.

Geologists have constructed a detailed history of sediment deposition that links particular bodies of rock in the crust of Earth to particular environments and processes. If petroleum geologists could find more oil and gas by interpreting the record of sedimentary rocks as having resulted from a single flood, they would cer-tainly favor the idea of such a flood, but they do not. Instead, these practical workers agree with academic geologists about the nature of depositional environments and geological time. Petroleum geologists have been pioneers in the recognition of fossil deposits that were formed over millions of years in such environments as meandering rivers, deltas, sandy barrier beaches, and coral reefs.

The example of petroleum geology demonstrates one of the great strengths of science. By using knowledge of the natural world to predict the consequences of our actions, science makes it possible to solve problems and create opportunities using technology. The detailed knowledge required to sustain our civilization could only have been derived through scientific investigation. The arguments of creationists are not driven by evidence that can be observed in the natural world. Special creation or supernatural intervention is not subjectable to meaningful tests, which require predicting plausible results and then checking these results through observation and experimentation. Indeed, claims of "special cre-ation" reverse the scientific process. The explanation is seen as unalterable, and evi-dence is sought only to support a particular conclusion by whatever means possible.

EVIDENCE SUPPORTING BIOLOGICAL EVOLUTION

Along path leads from the origins of primitive "life," which existed at least 3.5 billion years ago, to the profusion and diversity of life that exists today. This path is best understood as a product of evolution.

Contrary to popular opinion, neither the term nor the idea of biological evolution began with Charles Darwin and his foremost work, *On the Origin of Species by Means of Natural Selection* (1859). Many scholars from the ancient Greek philoso-phers on had inferred that similar species were descended from a common ancestor. The word "evolution" first appeared in the English language in 1647 in a nonbiolog-ical connection, and it became widely used in English for all sorts of progressions from simpler beginnings. The term Darwin most often used to refer to biological evolution was "descent with modification," which remains a good brief definition of the process today.

Darwin proposed that evolution could be explained by the differential survival of organisms following their naturally occurring variation—a process he termed "natural selection."

According to this view, the offspring of organisms differ from one another and from their parents in ways that are heritable—that is, they can pass on the differences genetically to their own offspring. Furthermore, organisms in nature typically produce more offspring than can survive and reproduce given the constraints of food, space, and other environmental resources. If a particular off-spring has traits that give it an advantage in a particular environment, that organism will be more likely to survive and pass on those traits. As differences accumulate over generations, populations of organisms diverge from their ancestors. Darwin's original hypothesis has undergone extensive modification and expan-sion, but the central concepts stand firm. Studies in genetics and molecular biology— fields unknown in Darwin's time—have explained the occurrence of the heredi-tary variations that are essential to natural selection. Genetic variations result from changes, or mutations, in the nucleotide sequence of DNA, the molecule that genes are made from. Such changes in DNAnow can be detected and described with great precision.

Genetic mutations arise by chance. They may or may not equip the organism with better means for surviving in its environment. But if a gene variant improves adaptation to the environment (for example, by allowing an organism to make better use of an available nutrient, or to escape predators more effectively—such as through stronger legs or disguising coloration), the organisms carrying that gene are more likely to survive and reproduce than those without it. Over time, their descendants will tend to increase, changing the average characteristics of the population. Although the genetic variation on which natural selection works is based on random or chance elements, natural selection itself produces "adaptive" change—the very opposite of chance.

Scientists also have gained an understanding of the processes by which new species originate. Anew species is one in which the individuals cannot mate and produce viable descendants with individuals of a preexisting species. The split of one species into two often starts because a group of individuals becomes geographically separated from the rest. This is particularly apparent in distant remote islands, such as the Galápagos and the Hawaiian archipelago, whose great distance from the Americas and Asia means that arriving colonizers will have little or no opportunity to mate with individuals remaining on those continents. Mountains, rivers, lakes, and other natural barriers also account for geographic separation between populations that once belonged to the same species.

Once isolated, geographically separated groups of individuals become genetically differentiated as a consequence of mutation and other processes, including natural selection. The origin of a species is often a gradual process, so that at first the repro-ductive isolation between separated groups of organisms is only partial, but it eventu-ally becomes complete. Scientists pay special attention to these intermediate situations, because they help to reconstruct the details of the process and to identify particular genes or sets of genes that account for the reproductive isolation between species. A particularly compelling example of speciation involves the 13 species of finches studied by Darwin on the Galápagos Islands, now known as Darwin's finches. The ancestors of these finches appear to have immigrated from the South American mainland to the Galápagos. Today the different species of finches on the island have distinct habitats, diets, and behaviors, but the mechanisms involved in speciation continue to operate. Aresearch group led by Peter and Rosemary Grant of Princeton University has shown that a single year of drought on the islands can drive evolutionary changes in the finches. Drought diminishes supplies of easily cracked nuts but permits the survival of plants that produce larger, tougher nuts. Droughts thus favor birds with strong, wide beaks that can break these tougher seeds, producing populations of birds with these traits. The Grants have estimated that if droughts occur about once every 10 years on the islands, a new species of finch might arise in only about 200 years.

The following sections consider several aspects of biological evolution in greater detail, looking at paleontology, comparative anatomy, biogeography, embryology, and molecular biology for further evidence supporting evolution.

The Fossil Record

Although it was Darwin, above all others, who first marshaled convincing evidence for biological evolution, earlier scholars had recognized that organisms on Earth had changed systematically over long periods of time. For example, in 1799 an engineer named William Smith reported that, in undisrupted layers of rock, fossils occurred in a definite sequential order, with more modern-appearing ones closer to the top. Because bottom layers of rock logically were laid down earlier and thus are older than top layers, the sequence of fossils also could be given a chronology from oldest to youngest. His findings were confirmed and extended in the 1830s by the paleontologist William Lonsdale, who recognized that fossil remains of organ-isms from lower strata were more primitive than the ones above. Today, many thousands of ancient rock deposits have been identified that show corresponding successions of fossil organisms.

Thus, the general sequence of fossils had already been recognized before Darwin conceived of descent with modification. But the paleontologists and geologists before Darwin used the sequence of fossils in rocks not as proof of biological evolution, but as a basis for working out the original sequence of rock strata that had been structurally disturbed by earthquakes and other forces. In Darwin's time, paleontology was still a rudimentary science. Large parts of the geological succession of stratified rocks were unknown or inadequately studied. Darwin, therefore, worried about the rarity of intermediate forms between some major groups of organisms.

Today, many of the gaps in the paleontological record have been filled by the research of paleontologists. Hundreds of thousands of fossil organisms, found in well-dated rock sequences, represent successions of forms through time and mani-fest many evolutionary transitions. As mentioned earlier, microbial life of the sim-plest type was already in existence 3.5 billion years ago. The oldest evidence of more complex organisms (that is, eucaryotic cells, which are more complex than bacteria) has been discovered in fossils sealed in rocks approximately 2 billion years old.

Multicellular organisms, which are the familiar fungi, plants, and animals, have been found only in younger geological strata. The following list presents the order in which increasingly complex forms of life appeared:

Life Form	Millions of Years Since First Known Appearance (Approximate)
Microbial (procaryotic cells)	3,500
Complex (eucaryotic cells)	2,000
First multicellular animals	670
Shell-bearing animals	540
Vertebrates (simple fishes)	490
Amphibians	350
Reptiles	310
Mammals	200
Nonhuman primates	60
Earliest apes	25
Australopithecine ancestors of humans	5
Modern humans	0.15 (150,000 years)

So many intermediate forms have been discovered between fish and amphib-ians, between amphibians and reptiles, between reptiles and mammals, and along the primate lines of descent that it often is difficult to identify categorically when the transition occurs from one to another particular species. Actually, nearly all fos-sils can be regarded as intermediates in some sense; they are life forms that come between the forms that preceded them and those that followed.

The fossil record thus provides consistent evidence of systematic change through time—of descent with modification. From this huge body of evidence, it can be predicted that no reversals will be found in future paleontological studies.

That is, amphibians will not appear before fishes, nor mammals before reptiles, and no complex life will occur in the geological record before the oldest eucaryotic cells. This prediction has been upheld by the evidence that has accumulated until now: no reversals have been found.

Common Structures

Inferences about common descent derived from paleontology are reinforced by comparative anatomy. For example, the skeletons of humans, mice, and bats are strikingly similar, despite the different ways of life of these animals and the diversi-ty of environments in which they flourish. The correspondence of these animals, bone by bone, can be observed in every part of the body, including the limbs; yet a person writes, a mouse runs, and a bat flies with structures built of bones that are different in detail but similar in general structure and relation to each other. Scientists call such structures homologies and have concluded that they are best explained by common descent. Comparative anatomists investigate such homologies, not only in bone structure but also in other parts of the body, working out rela-tionships from degrees of similarity. Their conclusions provide important inferences about the details of evolutionary history, inferences that can be tested by compar-isons with the sequence of ancestral forms in the paleontological record.

The mammalian ear and jaw are instances in which paleontology and compara-tive anatomy combine to show common ancestry through transitional stages. The lower jaws of mammals contain only one bone, whereas those of reptiles have sever-al. The other bones in the reptile jaw are homologous with bones now found in the mammalian ear. Paleontologists have discovered intermediate forms of mammal-like reptiles (Therapsida) with a double jaw joint—one composed of the bones that persist in mammalian jaws, the other consisting of bones that eventually became the hammer and anvil of the mammalian ear.

The Distribution of Species

Biogeography also has contributed evidence for descent from common ances-tors. The diversity of life is stupendous. Approximately 250,000 species of living plants, 100,000 species of fungi, and one million species of animals have been described and named, each occupying its own peculiar ecological setting or niche; and the census is far from complete. Some species, such as human beings and our companion the dog, can live under a wide range of environments. Others are amaz-ingly specialized. One species of a fungus (*Laboulbenia*) grows exclusively on the rear portion of the covering wings of a single species of beetle (*Aphaenops cronei*) found only in some caves of southern France.

The larvae of the fly *Drosophila car-cinophila* can develop only in specialized grooves beneath the flaps of the third pair of oral appendages of a land crab that is found only on certain Caribbean islands. How can we make intelligible the colossal diversity of living beings and the existence of such extraordinary, seemingly whimsical creatures as the fungus, beetle, and fly described above? And why are island groups like the Galápagos so often inhabited by forms similar to those on the nearest mainland but belonging to differ-ent species? Evolutionary theory explains that biological diversity results from the descendants of local or migrant predecessors becoming adapted to their diverse environments. This explanation can be tested by examining present species and local fossils to see whether they have similar structures, which would indicate how one is derived from the other. Also, there should be evidence that species without an established local ancestry had migrated into the locality.

Wherever such tests have been carried out, these conditions have been con-firmed. Agood example is provided by the mammalian populations of North and South America, where strikingly different native organisms evolved in isolation until the emergence of the isthmus of Panama approximately 3 million years ago. Thereafter, the armadillo, porcupine, and opossum—mammals of South American origin—migrated north, along with many other species of plants and animals, while the mountain lion and other North American species made their way across the isthmus to the south.

The evidence that Darwin found for the influence of geographical distribution on the evolution of organisms has become stronger with advancing knowledge. For example, approximately

2,000 species of flies belonging to the genus *Drosophila* are now found throughout the world. About one-quarter of them live only in Hawaii. More than a thousand species of snails and other land mollusks also are found only in Hawaii. The biological explanation for the multiplicity of related species in remote localities is that such great diversity is a consequence of their evolution from a few common ancestors that colonized an isolated environment. The Hawaiian Islands are far from any mainland or other islands, and on the basis of geological evidence they never have been attached to other lands. Thus, the few colonizers that reached the Hawaiian Islands found many available ecological niches, where they could, over numerous generations, undergo evolutionary change and diversifi-cation. No mammals other than one bat species lived in the Hawaiian Islands when the first human settlers arrived; similarly, many other kinds of plants and animals were absent.

The Hawaiian Islands are not less hospitable than other parts of the world for the absent species. For example, pigs and goats have multiplied in the wild in Hawaii, and other domestic animals also thrive there. The scientific explanation for the absence of many kinds of organisms, and the great multiplication of a few kinds, is that many sorts of organisms never reached the islands, because of their geo-graphic isolation. Those that did reach the islands diversified over time because of the absence of related organisms that would compete for resources.

Similarities During Development

Embryology, the study of biological development from the time of conception, is another source of independent evidence for common descent. Barnacles, for instance, are sedentary crustaceans with little apparent similarity to such other crustaceans as lobsters, shrimps, or copepods. Yet barnacles pass through a free-swimming larval stage in which they look like other crustacean larvae. The similar-ity of larval stages supports the conclusion that all crustaceans have homologous parts and a common ancestry.

Similarly, a wide variety of organisms from fruit flies to worms to mice to humans have very similar sequences of genes that are active early in development. These genes influence body segmentation or orientation in all these diverse groups. The presence of such similar genes doing similar things across such a wide range of organisms is best explained by their having been present in a very early common ancestor of all of these groups.

New Evidence from Molecular Biology

The unifying principle of common descent that emerges from all the foregoing lines of evidence is being reinforced by the discoveries of modern biochemistry and molecular biology.

The code used to translate nucleotide sequences into amino acid sequences is essentially the same in all organisms. Moreover, proteins in all organisms are invariably composed of the same set of 20 amino acids. This unity of composition and function is a powerful argument in favor of the common descent of the most diverse organisms.

In 1959, scientists at Cambridge University in the United Kingdom determined the three-dimensional structures of two proteins that are found in almost every mul-ticelled animal: hemoglobin and myoglobin. Hemoglobin is the protein that carries oxygen in the blood. Myoglobin receives oxygen from hemoglobin and stores it in the tissues until needed. These were the first three-dimensional protein structures to be solved, and they yielded some key insights. Myoglobin has a single chain of 153 amino acids wrapped around a group of iron and other atoms (called "heme") to which oxy-gen binds. Hemoglobin, in contrast, is made of up four chains: two identical chains consisting of 141 amino acids, and two other identical chains consisting of 146 amino acids. However, each chain has a heme exactly like that of myoglo-bin, and each of the four chains in the hemoglobin molecule is folded exactly like myoglobin.

It was immediately obvious in 1959 that the two molecules are very closely related. During the next two decades, myoglobin and hemoglobin sequences were determined for dozens of mammals, birds, reptiles, amphibians, fish, worms, and molluscs. All of these sequences were so obviously related that they could be com-pared with confidence with the three-dimensional structures of two selected standards—whale myoglobin and horse hemoglobin. Even more significantly, the differences between sequences from different organisms could be used to con-struct a family tree of hemoglobin and myoglobin variation among organisms. This tree agreed completely with observa-tions derived from paleontology and anatomy about the common descent of the cor-responding organisms.

Similar family histories have been obtained from the three-dimensional struc-tures and amino acid sequences of other proteins, such as cytochrome *c* (a protein engaged in energy transfer) and the digestive proteins trypsin and chymotrypsin. The examination of molecular structure offers a new and extremely powerful tool for studying evolutionary relationships. The quantity of information is potentially huge—as large as the thousands of different proteins contained in living organisms, and limited only by the time and resources of molecular biologists.

As the ability to sequence the nucleotides making up DNA has improved, it also has become possible to use genes to reconstruct the evolutionary history of organ-isms. Because of mutations, the sequence of nucleotides in a gene gradually changes over time. The more closely related two organisms are, the less different their DNA will be. Because there are tens of thousands of genes in humans and other organisms, DNAcontains a tremendous amount of information about the evo-lutionary history of each organism.

Genes evolve at different rates because, although mutation is a random event, some proteins are much more tolerant of changes in their amino acid sequence than are other proteins. For this reason, the genes that encode these more tolerant, less constrained proteins evolve faster. The average rate at which a particular kind of gene or protein evolves gives rise to the concept of a "molecular clock." Molecular clocks run rapidly for less constrained proteins and slowly for more constrained proteins, though they all time the same evolutionary events. The figure on this page compares three molecular clocks: for cytochrome *c* pro-teins, which interact intimately with other macromolecules and are quite con-strained in their amino acid sequences; for the less rigidly constrained hemoglobins, which interact mainly with oxygen and other small molecules; and for fibrinopeptides, which are protein fragments that are cut from larger proteins (fibrinogens) when blood clots. The clock for fibrinopeptides runs rapidly; 1 percent of the amino acids change in a little longer than 1 million years. At the other extreme, the molecu-lar clock runs slowly for cytochrome *c*; a 1 percent change in amino acid sequence requires 20 million years. The hemoglobin clock is intermediate.

The concept of a molecular clock is useful for two purposes. It determines evolutionary relationships among organisms, and it indicates the time in the past when species started to diverge from one another. Once the clock for a particular gene or protein has been calibrated by reference to some event whose time is known, the actual chronological time when all other events occurred can be determined by examining the protein or gene tree.

An interesting additional line of evidence supporting evolution involves sequences of DNA known as "pseudogenes." Pseudogenes are remnants of genes that no longer function but continue to be carried along in DNA as excess baggage. Pseudogenes also change through time, as they are passed on from ancestors to descendants, and they offer an especially useful way of reconstructing evolutionary relationships.

With functioning genes, one possible explanation for the relative similarity between genes from different organisms is that their ways of life are similar—for example, the genes from a horse and a zebra could be more similar because of their similar habitats and behaviors than the genes from a horse and a tiger. But this pos-sible explanation does not work for pseudogenes, since they perform no function. Rather, the degree of similarity between pseudogenes must simply reflect their evo-lutionary relatedness. The more remote the last common ancestor of two organisms, the more dissimilar their pseudogenes will be.

The evidence for evolution from molecular biology is overwhelming and is growing quickly. In some cases, this molecular evidence makes it possible to go beyond the paleontological evidence. For example, it has long been postulated that whales descended from land mammals that had returned to the sea. From anatomi-cal and paleontological evidence, the whales' closest living land relatives seemed to be the even-toed hoofed mammals (modern cattle, sheep, camels, goats, etc.). Recent comparisons of some milk protein genes (beta-casein and kappa-casein) have confirmed this relationship and have suggested that the closest land-bound living relative of whales may be the hippopotamus. In this case, molecular biology has augmented the fossil record.

Creationism and the Evidence for Evolution

Some creationists cite what they say is an incomplete fossil record as evidence for the failure of evolutionary theory. The fossil record was incomplete in Darwin's time, but many of the important gaps that existed then have been filled by subsequent paleontological research. Perhaps the most persuasive fossil evidence for evo-lution is the consistency of the sequence of fossils from early to recent. Nowhere on Earth do we find, for example, mammals in Devonian (the age of fishes) strata, or human fossils coexisting with dinosaur remains. Undisturbed strata with simple unicellular organisms predate those with multicellular organisms, and invertebrates precede vertebrates; nowhere has this sequence been found inverted. Fossils from adjacent strata are more similar than fossils from temporally distant strata. The most reasonable scientific conclusion that can be drawn from the fossil record is that descent with modification has taken place as stated in evolutionary theory. Special creationists argue that "no one has seen evolution occur." This misses the point about how science tests hypotheses. We don't see Earth going around the sun or the atoms that make up matter. We "see" their consequences. Scientists infer that atoms exist and Earth revolves because they have tested predictions derived from these concepts by extensive observation and experimentation.

Furthermore, on a minor scale, we "experience" evolution occurring every day. The annual changes in influenza viruses and the emergence of antibiotic-resistant bacteria are both products of evolutionary forces. Indeed, the rapidity with which organisms with short generation times, such as bacteria and viruses, can evolve under the influence of their environments is of great medical significance. Many laboratory experiments have shown that, because cf mutation and natural selection, such microorganisms can change in specific ways from those of immediately pre-ceding generations.

On a larger scale, the evolution of mosquitoes resistant to insecticides is another example of the tenacity and adaptability of organisms under environmental stress. Similarly, malaria parasites have become resistant to the drugs that were used extensively to combat them for many years. As a consequence, malaria is on the increase, with more than 300 million clinical cases of malaria occurring every year. Molecular evolutionary data counter a recent proposition called "intelligent design theory." Proponents of this idea argue that structural complexity is proof of the direct hand of God in specially creating organisms as they are today. These arguments echo those of the 18th century cleric William Paley who held that the vertebrate eye, because of its intricate organization, had been specially designed in its present form by an omnipotent Creator. Modern-day intelligent design proponents argue that molecular structures such as DNA, or molecular processes such as the many steps that blood goes through when it clots, are so irreducibly complex that they can function only if all the components are operative at once. Thus, pro-ponents of intelligent design say that these structures and processes could not have evolved in the stepwise mode characteristic of natural selection.

However, structures and processes that are claimed to be "irreducibly" complex typically are not on closer inspection. For example, it is incorrect to assume that a complex structure or biochemical process can function only if all its components are present and functioning as we see them today. Complex biochemical systems can be built up from simpler systems through natural selection. Thus, the "history" of a protein can be traced through simpler organisms. Jawless fish have a simpler hemoglobin than do jawed fish, which in turn have a simpler hemoglobin than mammals. The evolution of complex molecular systems can occur in several ways. Natural selection can bring together parts of a system for one function at one time and then, at a later time, recombine those parts with other systems of components to produce a system that has a different function. Genes can be duplicated, altered, and then amplified through natural selection. The complex biochemical cascade resulting in blood clotting has been explained in this fashion.

Similarly, evolutionary mechanisms are capable of explaining the origin of highly complex anatomical structures. For example, eyes may have evolved independently many times during the history of life on Earth. The steps proceed from a simple eye spot made up of light-sensitive retinula cells (as is now found in the flatworm), to for-mation of individual photosensitive units (ommatidia) in insects with light focusing lenses, to the eventual formation of an eye with a single lens focusing images onto a retina. In humans and other vertebrates, the retina consists not only of photoreceptor cells but also of several types of neurons that begin to analyze the visual image. Through such gradual steps, very different kinds of eyes have evolved, from simple light-sensing organs to highly complex systems for vision.

HUMAN EVOLUTION

Studies in evolutionary biology have led to the conclusion that human beings arose from ancestral primates. This association was hotly debated among sci-entists in Darwin's day. But today there is no significant scientific doubt about the close evolutionary relationships among all primates, including humans. Many of the most important advances in paleontology over the past century relate to the evolutionary history of humans. Not one but many connecting links—intermediate between and along various branches of the human family tree—have been found as fossils. These linking fossils occur in geological deposits of intermediate age. They document the time and rate at which primate and human evolution occurred.

Scientists have unearthed thousands of fossil specimens representing members of the human family. Agreat number of these cannot be assigned to the modern human species, *Homo sapiens*. Most of these specimens have been well dated, often by means of radiometric techniques.

They reveal a well-branched tree, parts of which trace a general evolutionary sequence leading from ape-like forms to modern humans. Paleontologists have discovered numerous species of extinct apes in rock strata that are older than four million years, but never a member of the human family at that great age. *Australopithecus*, whose earliest known fossils are about four million years old, is a genus with some features closer to apes and some closer to modern humans. In brain size, *Australopithecus* was barely more advanced than apes.

A number of features, including long arms, short legs, intermediate toe structure, and features of the upper limb, indicate that the members of this species spent part of the time in trees. But they also walked upright on the ground, like humans. Bipedal tracks of *Australopithecus* have been discovered, beautifully preserved with those of other extinct animals, in hardened volcanic ash. Most of our *Australopithecus* ancestors died out close to two-and-a-half million years ago, while other *Australopithecus* species, which were on side branches of the human tree, survived alongside more advanced hominids for another million years.

Distinctive bones of the oldest species of the human genus, *Homo*, date back to rock strata about 2.4 million years old. Physical anthropologists agree that *Homo* evolved from one of the species of *Australopithecus*. By two million years ago, early members of *Homo* had an average brain size one-and-a-half times larger than that of *Australopithecus*, though still substantially smaller than that of modern humans. The shapes of the pelvic and leg bones suggest that these early *Homo* were not part-time climbers like *Australopithecus* but walked and ran on long legs, as modern humans do. Just as *Australopithecus* showed a complex of ape-like, human-like, and intermediate features, so was early *Homo* intermediate between *Australopithecus* and modern humans in some features, and close to modern humans in other respects. The earliest stone tools are of virtually the same age as the earliest fossils of *Homo*. Early *Homo*, with its larger brain than *Australopithecus*, was a maker of stone tools.

The fossil record for the interval between 2.4 million years ago and the present includes the skeletal remains of several species assigned to the genus *Homo*. The more recent species had larger brains than the older ones. This fossil record is com-plete enough to show that the human genus first spread from its place of origin in Africa to Europe and Asia a little less than two million years ago. Distinctive types of stone tools are associated with various populations. More recent species with larger brains generally used more sophisticated tools than more ancient species. Molecular biology also has provided strong evidence of the close relationship between humans and apes. Analysis of many proteins and genes has shown that humans are genetically similar to chimpanzees and gorillas and less similar to orangutans and other primates.

DNA has even been extracted from a well-preserved skeleton of the extinct human creature known as Neanderthal, a member of the genus *Homo* and often considered either as a subspecies of *Homo sapiens* or as a separate species. Application of the molecular clock, which makes use of known rates of genetic mutation, suggests that Neanderthal's lineage diverged from that of modern *Homo sapiens* less than half a million years ago, which is entirely compatible with evidence from the fossil record. Based on molecular and genetic data, evolutionists favor the hypothesis that modern *Homo sapiens*, individuals very much like us, evolved from more archaic humans about 100,000 to 150,000 years ago. They also believe that this transition occurred in Africa, with modern humans then dispersing to Asia, Europe, and even-tually Australasia and the Americas.

Discoveries of hominid remains during the past three decades in East and South Africa, the Middle East, and elsewhere have combined with advances in molecular biology to initiate a new discipline—molecular paleoanthropology. This field of inquiry is providing an ever-growing inventory of evidence for a genetic affinity between human beings and the African apes. Opinion polls show that many people believe that divine intervention actively guided the evolution of human beings. Science cannot comment on the role that supernatural forces might play in human affairs. But scientific investigations have concluded that the same forces responsible for the evolution of all other life forms on Earth can account for the evolution of human beings.

CONCLUSION

Science is not the only way of acquiring knowledge about ourselves and the world around us. Humans gain understanding in many other ways, such as through literature, the arts, philosophical reflection, and religious experience. Scientific knowledge may enrich aesthetic and moral perceptions, but these subjects extend beyond science's realm, which is to obtain a better understanding of the nat-ural world.

The claim that equity demands balanced treatment of evolutionary theory and special creation in science classrooms reflects a misunderstanding of what science is and how it is conducted. Scientific investigators seek to understand natural phenomena by observation and experimentation. Scientific interpretations of facts and the explanations that account for them therefore must be testable by observation and experimentation.

Creationism, intelligent design, and other claims of supernatural intervention in the origin of life or of species are not science because they are not testable by the methods of science. These claims subordinate observed data to statements based on authority, revelation, or religious belief. Documentation offered in support of these claims is typically limited to the special publications of their advocates. These publications do not offer hypotheses subject to change in light of new data, new interpretations, or demonstration of error. This contrasts with science, where any hypothesis or theory always remains subject to the possibility of rejection or modifi-cation in the light of new knowledge. No body of beliefs that has its origin in doctrinal material rather than scientific observation, interpretation, and experimentation should be admissible as science in any science course. Incorporating the teaching of such doctrines into a science curriculum compromises the objectives of public education. Science has been greatly successful at explaining natural processes, and this has led not only to increased understanding of the universe but also to major improvements in technology and public health and welfare. The growing role that science plays in modern life requires that science, and not religion, be taught in science classes.

Evolution in the broadest sense explains that what we see today is different from what existed in the past. Galaxies, stars, the solar system, and Earth have changed through time, and so has life on Earth. Biological evolution concerns changes in living things during the history of life on Earth. It explains that living things share common ancestors. Over time, biologi-cal processes such as natural selection give rise to new species. Darwin called this process "descent with modification," which remains a good definition of biological evolution today.

Isn't Evolution just an Inference?

No one saw the evolution of one-toed horses from three-toed horses, but that does not mean that we cannot be confident that horses evolved. Science is practiced in many ways besides direct observation and experimentation. Much scientific dis-covery is done through indirect experimentation and observation in which infer-ences are made, and hypotheses generated from those inferences are tested. For instance, particle physicists cannot directly observe subatomic particles because the particles are too small. They make inferences about the weight, speed, and other properties of the particles based on other observations. Alogical hypoth-esis might be something like this: If the weight of this particle is *Y*, when I bombard it, *X* will happen. If *X* does not happen, then the hypothesis is disproved. Thus, we can learn about the natural world even if we cannot directly observe a phenome-non—and that is true about the past, too.

In historical sciences like astronomy, geology, evolutionary biology, and archae-ology, logical inferences are made and then tested against data. Sometimes the test cannot be made until new data are available, but a great deal has been done to help us understand the past. For example, scorpionflies (*Mecoptera*) and true flies (*Diptera*) have enough similarities that entomologists consider them to be closely related. Scorpionflies have four wings of about the same size, and true flies have a large front pair of wings but the back pair is replaced by small club-shaped structures. If two-winged flies evolved from scorpionfly-like ancestors, as comparative anatomy suggests, then an intermediate true fly with four wings should have existed— and in 1976 fossils of such a fly were discovered. Furthermore, geneticists have found that the number of wings in flies can be changed through mutations in a single gene. Something that happened in the past is thus not "off limits" for scientific study.

Hypotheses can be made about such phenomena, and these hypotheses can be tested and can lead to solid conclusions. Furthermore, many key mechanisms of evolu-tion occur over relatively short periods and can be observed directly—such as the evolution of bacteria resistant to antibiotics.

Evolution is a well-supported theory drawn from a variety of sources of data, including observations about the fossil record, genetic information, the distribution of plants and animals, and the similarities across species of anatomy and develop-ment. Scientists have inferred that descent with modification offers the best scientif-ic explanation for these observations.

Is Evolution a Fact or a Theory?

The theory of evolution explains how life on Earth has changed. In scientific terms, "theory" does not mean "guess" or "hunch" as it does in everyday usage. Scientific theories are explanations of natural phenomena built up logically from testable observations and hypotheses. Biological evolution is the best scientific explanation we have for the enormous range of observations about the living world. Scientists most often use the word "fact" to describe an observation. But scien-tists can also use fact to mean something that has been tested or observed so many times that there is no longer a compelling reason to keep testing or looking for examples. The occurrence of evolution in this sense is a fact.

Scientists no longer question whether descent with modification occurred because the evidence support-ing the idea is so strong.

Don't many Famous Scientists Reject Evolution?

No. The scientific consensus around evolution is overwhelming. Those opposed to the teaching of evolution sometimes use quotations from prominent sci-entists out of context to claim that scientists do not support evolution. However, examination of the quotations reveals that the scientists are actually disputing some aspect of *how* evolution occurs, not *whether* evolution occurred.

For example, the biologist Stephen Jay Gould once wrote that "the extreme rarity of transitional forms in the fossil record persists as the trade secret of paleontology." But Gould, an accomplished paleontologist and eloquent educator about evolution, was arguing about *how* evolution takes place. He was discussing whether the rate of change of species is slow and gradual or whether it takes place in bursts after long periods when little change occurs—an idea known as punctuated equilibrium. As Gould writes in response, "This quotation, although accurate as a partial citation, is dis-honest in leaving out the following explanatory material showing my true pur-pose—to discuss rates of evolutionary change, not to deny the fact of evolution itself." Gould defines punctuated equilibrium as follows:

Punctuated equilibrium is neither a creationist idea nor even a non-Darwinian evolutionary theory about sudden change that produces a new species all at once in a single generation. Punctuated equilibrium accepts the conventional idea that new species form over hundreds or thousands of generations and through an extensive series of intermediate stages. But geological time is so long that even a few thousand years may appear as a mere "moment" relative to the several million years of exis-tence for most species. Thus, rates of evolution vary enormously and new species may appear to arise "suddenly" in geological time, even though the time involved would seem long, and the change very slow, when compared to a human lifetime.

If Humans Evolved from Apes, why are there Still Apes?

Humans did not evolve from modern apes, but humans and modern apes shared a common ancestor, a species that no longer exists. Because we share a recent common ancestor with chimpanzees and gorillas, we have many anatomical, genetic, biochemical, and even behavioral similarities with these African great apes. We are less similar to the Asian apes—orangutans and gibbons—and even less simi-lar to monkeys, because we share common ancestors with these groups in the more distant past.

Evolution is a branching or splitting process in which populations split off from one another and gradually become different. As the two groups become isolated from each other, they stop sharing genes, and eventually genetic differences increase until members of the groups can no longer interbreed. At this point, they have become separate species. Through time, these two species might give rise to new species, and so on through millennia.

Why can't we Teach Creation Science in my School?

The courts have ruled that "creation science" is actually a religious view. Because public schools must be religiously neutral under the U.S. Constitution, the courts have held that it is unconstitutional to present creation science as legitimate scholarship.

In particular, in a trial in which supporters of creation sciencetestified in sup-port of their view, a district court declared that creation science does not meet the tenets of science as scientists use the term (*McLean v. Arkansas Board of Education*). The Supreme Court has held that it is illegal to require that creation science be taught when evolution is taught (*Edwards v. Aguillard*).

In addition, district courts have decided that individual teachers cannot advocate creation science on their own (*Peloza v. San Juan Capistrano School District* and *Webster v. New Lennox School District*). (See *Teaching About Evolution and the Nature of Science*, Appendix A. National Academy of Sciences, Washington, D.C. 1998.)

Teachers' organizations such as the National Science Teachers Association, the National Association of Biology Teachers, the National Science Education Leadership Association, and many others also have rejected the science and peda-gogy of creation science and have strongly discouraged its presentation in the public schools. In addition, a coalition of religious and other organizations has noted in "A Joint Statement of Current Law" that "in science class, [schools] may present only genuinely scientific critiques of, or evidence for, any explanation of life on Earth, but not religious critiques (beliefs unverifiable by scientific methodology)." (See *Teaching About Evolution and the Nature of Science*, Appendices B and C, National Academy of Sciences, Washington, D.C., 1998.)

Some argue that "fairness" demands the teaching of creationism along with evolution. But a science curriculum should cover science, not the religious views of particular groups or individuals.

If Evolution is Taught in Schools, shouldn't Creationism be given Equal Time?

Some religious groups deny that microorganisms cause disease, but the science curriculum should not therefore be altered to reflect this belief. Most people agree that students should be exposed to the best possible scholarship in each field. That scholarship is evaluated by professionals and educators in those fields. Scientists as well as educators have concluded that evolution—and only evolution—should be taught in science classes because it is the only *scientific* explanation for why the universe is the way it is today.

Many people say that they want their children to be exposed to creationism in school, but there are thousands of different ideas about creation among the world's people. Comparative religions might comprise a worthwhile field of study, but not one appropriate for a science class. Furthermore, the U.S. Constitution states that schools must be religiously neutral, so legally a teacher cannot present any particu-lar creationist view as being more "true" than others.

6

BIOLOGY AND BIOTECHNOLOGY

The protein insulin, products by the islet cells in the pancreas, is required to regulate the uptake of glucose from the blood into diverse cells in the body. Diabetics have too much glucose (a sugar generated or contained in food) in the blood and not enough in the high-energy utilizing cell types.

Excess glucose can result from one of two factors:

1. *Low Insulin Production*: the islet cells die or become less productive. This is typical of juvenile diabetes. This type of diabetes can be treated by administering insulin: without insulin the patient will enter a diabetic coma and die.
2. *Body Cells become Resistant to Insulin*: the pancreas produces the normal amount of insulin protein, but other cells fail to recognize insulin and hence fail to take up sufficient glucose. This is typical of adult onset diabetes. This type of diabetes cannot be treated with extra insulin--the best course is lowering the amount of glucose in the diet and increasing exercise to make cells more likely to take up sugar molecules. With proper management, this type of diabetes can often be controlled so that there are no disease symptoms.

Who is Susceptible?

There is a strong genetic component to juvenile onset diabetes. Adult onset diabetes is usually caused by a poor diet and too little exercise.

Before recombinant DNA, what was the source of insulin?

Cadavers and animal pancreas products were the primary sources. Both sources pose a risk of allergenic reaction, because insulin had to be purified from a complex mixture of proteins. There is also the potential for transmission of disease, particularly viruses and especially of "unknown" viruses for which there is no test (yet). Recall that hemophiliacs often contracted AIDS in the 1980s, because their disease required treatment with blood proteins concentrated from ~10,000 blood donors per dose. Once the incidence of AIDS reached a certain level, it was highly likely that their medicine contained the HIV (Human Immune Deficiency Virus) given the large number of blood donors required to generate one dose of medicine.

By the mid-1980s recombinant DNA tests were devised to screen for HIV in blood samples: these blood samples were not used in transfusions or to prepare blood products for medicine. Today, however, hemophiliacs like diabetics who need insulin, receive recombinant protein produced in bacteria.

This source is of higher purity and safety than the material purified from human or animal sources. Bacteria and humans do not share the same viruses, so there is little chance of infection. Large batches of insulin can be produced in fermentation tanks, and each batch can be tested for purity and efficacy in standard assays. The standardization of dosage is a tremendous benefit to those taking insulin.

How was the human gene for insulin cloned?

1. A portion of the sequence of amino acids in the insulin protein was already known, because it is a small, intensively studied protein
2. Central dogma of biology is accurate but does not provide enough specificity.

   DNA encodes messenger RNA that specifies triplet codons that determine the order of amino acids in a protein.

   However, there are multiple codons for some amino acids, therefore, the order of amino acids does not predict "exactly" the sequence of the DNA.

3. Antibodies are proteins that recognize their "antigen" very specifically. Purified human insulin protein injected into mice or rabbits is an antigen, and the animal will produce antibodies that recognize the "foreign protein." This principle is key to vaccination, in which you are exposed to a protein from a disease organism and your body generates antibodies that will bind and guide the destruction of the foreign protein (and the disease organism).
4. When messenger RNA is translated into protein on the cellular ribosomes, the nascent chain of amino acids is also present. If the protein synthesizing machinery is partially purified, antibodies to insulin can be used to precipitate the ribosomes translating the insulin mRNA.
5. Reverse transcriptase, a protein RNA viruses used to make cDNA (complementary DNA) copies of themselves, can be used in a test tube to make a cDNA copy of the insulin mRNA.
6. The product is a single stranded DNA molecule; if DNA polymerase enzyme is added, then it copies the template single strand to synthesize the complementary strand using a simple formula: G specifies C, A specifies T, C specifies G, and T specifies A.
7. The double stranded cDNA had "linkers" of known DNA composition attached to its ends using DNA ligase, an enzyme that attaches pieces of DNA together.
8. A purified, recipient plasmid was digested with a restriction enyzme to change the circular plasmid into a linear piece of DNA with "sticky ends." The same enzyme was used to digest the cDNA linkers creating similar "sticky ends." Sticky ends are overhanging, single stranded DNA bases such as GATC in the diagram at the end of a double stranded DNA molecule. This end will bind to a complementary sticky end CTAG to reform a duplex DNA molecule. After the bases pair, DNA ligase is used to seal up the connections making an intact double stranded molecule.

Double stranded G A T C Double stranded
Double stranded C T A G Double stranded

9. The insulin cDNA molecule was sequenced using the "new" DNA sequencing technology and it matched the fragmentary pieces of protein sequence information.
10. The insulin cDNA in a bacterial plasmid was returned to E. coli bacteria where it could be expressed. This step is called transformation. Antibody to human insulin reacted with the transformed bacteria but not with control bacteria lacking the recombinant plasmid.
11. Insulin protein purified from transformed E. coli successfully treated diabetic mice.

Something to think about:

Advocates for juvenile diabetes research want to preserve the ability of scientists to do research on embryonic stem cells. Stem cells can be coaxed to differentiate into many different cell types during lab manipulation. There is "hope" that stem cells could be used to replace the defective Islet cells in the pancreas of juvenile diabetics. There are millions of children with this disease. Taking insulin shots is a maintenance medicine, not a cure. Yet, many Americans are opposed--some very strongly opposed--to the use of embryo cells in research and treatment programs. What do you think about these issues? We'll be discussing stem cells later in the course, along with animal cloning and the use of animals to grow "replacement parts" for humans.

Biotechnology in microbes, plants and animals. Three snapshots.

BACTERIA-WHERE IT ALL BEGAN

What is ice nucleation and why do we care?

Frost! Frost causes and estimated $1 billion damage to crops in the United States. So frost reduction is very attractive to the farmer and already big business. Farmers have approached this by smug pots on cold nights and fans to move the warm air, covering plants, and additional insulation to plant roots or try to eliminate the ice-nucleation bacteria! Here is where genetic engineering came into play. Spraying the plants with bacteria that inhibit the growth of ice-nucleation bacteria. The bacterium primarily responsible for ice-nucleation is Pseudomonas syringae. A protein found on the outer cell wall of the bacterium serves as a nucleus for ice crystals to form when temperatures drop below 0 degrees.

Steve Lindow at UC Berkely showed that if P syringae is genetically altered to remove the gene for ice-nucleation then it loses its ability to result in freezing at this relatively "high" temperature. These genetically altered bacteria were referred to as ice minus bacteria.

How did Steve Lindow make this mutant bacterium?

Now the controversy-The company Advanced Genetic Sciences, known as AGS applied to the government to do field trials to see whether ice-minus Pseudomonas would result in crops that suffered less frost damage. Keep in mind that sufficient numbers of the ice-minus bacterium must be present to outcompete the wild type Pseudomonas. Despite this understanding in 1983 the NIH, through the Recombinant DNA Advisory Committee approved the field trials for the ice-minus Pseudomonas on a strawberry field. Protest and lengthy legal battles ensued. One of the main arguments in the legal battle concerned the possible role these bacteria may play in the weather. No ecologist, botanist or population geneticists were on the NIH committee. The initial case AGS lost, but finally in 1987 field trials did take place.

Was it worth it? Was this the right outcome?

In this case scientist could not agree on the right answer, microbiologist could argue that better strains of Rhizobium, nitrogen-fixing bacteria have been deliberately introduced into the environment, for decades while ecologist could argue the unknown effect that a change in the microflora might have.

Would this experiment cause this controversy today?

Changes in US regulations allow for a much broader range of simple field trials that in the 1980. Are we OK with that?

Plants-Can we really just eat a tomato instead of a vaccine?

Why do our tomatoes look so beautiful and taste so awful?

Tomatoes too are big business, amounting to over $3.5 billion. Yet we have all had the experience of buying tomatoes that look great but have no taste and weird texture. These tomatoes have usually been picked before they are ripe (i.e. green), and "ripened" artificially with ethylene to look red and beautiful. Since the tomato has been separated from the plant before ripening even though it looks red and delicious no additionally sugar and other flavors have been pumped up into the tomato. The good news is these hard as rocks tomatoes are easy to ship and unlikely to be damaged in transit.

Would it be possible to make a vine ripened tomato that could easily be shipped?

Several companies decided to tackle this question including Calgene, Monsanto and Zeneca Seeds. Two approaches were tried.

a. Use a bacterial gene to prevent ethylene from forming in the tomato allowing for slower ripening on the vine. A Pseudomonas gene has been used in this way and these tomatoes can be picked later and do stay firm longer.

b. Use an "anti-sense RNA molecule to block the normal "sense" mRNA of a protein associated with fruit ripening. This approach was used to prevent the synthesis of the enzyme polygalacturonase-PG for short. PG works in a tomato when it goes from green to red, resulting in the bread down of some of the cell wall components, to make the fruit softer. The good news is PG doesn't have any affect on ripening so the result is a sweet flavorful tomato. (Nt.-Sometimes tomato paste manufactures often heated tomatoes to inactivate PG to keep the tomatoes firm.) Calgene did this experiment and created the FLAVR SAVRTM tomato, commercially know as the McGreggor tomato.

Are the FLAVR SAVRTM tomatoes really genetically engineered is it really that different from tomatoes that have been bred for other features? What happens if an antibiotic was used as a marker on the plant transformation. What happens to the vitamins?

AMIMALS-COMPLEX ETHICAL AND MORAL ISSUES

Before there was Dolly there was Tracy. Emphysema is a disease where a person's alveoli degernerate and carbon dioxide and oxygen cannot be exchanged. Damage is perrmanent and the most common cause is cigarette smoking, but asthma, chronic bronchitis and stuff in the air can also play a role. However, there is also a genetic disease that effects about one in every few thousand people. These people lack a-antitrypsin (AAT).

The enzyme Trypsin is normally released by the white blood cells and serves as part of the bodies normal defence against infection by chewing up microbial proteins. AAT protects the lungs from attack from these powerful enzymes. Treatment for this form of disease was to replace the missing AAT with AAT purified from human blood. An average patient needed about 200grams/year needless to say this was difficult and very expensive.

In 1990, scientist at PPL Therapeutics in Edinburgh transfeered copies of the human AAT gene in to five sheep embryos. Four of the sheep were female and the sheep named Tracy produced 35 grams o fAAT in each litre of here milk!!

Why didn't scientist just clone into bacteria (like insulin) or plants. AAT, like many human enzymes, are required to be modified after the protein is read off the RNA. In the case of AAT, the protein must be glycosylated whree a sugar molecule is added. In sheep this post transcriptional modification is completed to the AAT is active in humans.

The PPL Therapeutics scientist also added control regions to the AAT gene they gave to Tracy to ensure that the gene was only expressed in the mammary glands so there is no danger to the sheep where the gene might be expressed in nerve, blood or other tissue-only milk. Presently mice, rabbits, goats and cattlehave been engineered to produce a variety of proteins. In fact AAT was first produced in mice as a model organism to understand whether the experiment could be done in a mammal and whether tne post tranlational prosessing was the same.

1. Somatic body is diploid: each cells contains one complete set of chromosomes from the mother and father. Total number of chromosomes = 46 in human. These chromosomes are numbered; you have a chromosome #1 inherited from your mother and a chromosome #1 inherited from your father.
2. Gametes are produced in sex-specific reproductive organs. Eggs and sperm are haploid, because they contain just one of each type of chromosome. Total number of chromosomes = 23.
3. Some genetic diseases occur because of a chromosome abnormality: 3 copies of chromosome 21 results in Down's syndrome.
4. Sex determination in mammals depends on the X and Y chromosomes. XX is female and XY is male. For this reason, the sex of children entirely depends on whether a sperm contains an X or Y chromosome; the mother always transmits an X chromosome to her offspring.

   Coin flipping problem

   What are the "odds" of having a boy? A girl?

   If a family has 3 daughters, what are the odds that the next child is male?

   Equation: (1/2)n = odds of having "n" children of the same sex

   What fraction of families will have

   2 daughters, no sons

   5 sons, no daughters
5. The "maternal" and "paternal" chromosomes assort independently during gamete production. This means that each of your gametes has a subset of chromosome from each of your parents. This is one of MENDEL'S LAWS.

6. To increase "diversity" chromosomes also recombine during gamete production. Meiosis is the process of reducing the number of chromosomes from 46 to 23 in each gamete. When the two chromosome #1s are paired during meiosis, they can exchange materials. This makes "hybrid" chromosomes that contain some genes from the maternal chromosome and some from the paternal. Genetic recombination is very carefully regulated in the cell so that no DNA is lost! Precise recombination exchanges occur between the chromosomes.
7. Genetic recombination makes it impossible for you to sire an "exact" duplicate of yourself--not only do you need a sexual partner, but none of your gametes contains exactly the same information you have in your somatic cells. Biological cloning--growing an intact organism from a somatic cell or nucleus (set of chromosomes)--is the only method we have for making an exact duplicate of a complex organism.
8. Nonetheless, children display many of the traits (phenotypes) present in their parents, grandparents, and more distant relatives.
9. Each gene in the human body is found at a specific chromosome location. Because you have two chromosomes of each type, you have two copies of each gene (diploidy).
10. Genes have multiple alleles--variations on that genes. A good analogy is the colors of the rainbow. A gene for "color" could be occupied by an allele for any particular color.
11. Some alleles are DOMINANT and some are recessive.

The combination of alleles determines the phenotype. Imagine the genetic disease "alpha" produced when a person has the a//a genotype. If the Father is A//a and the Mother is A//a also, they can produce four kinds of children. The outcome is conveniently represented in a Punnett Square where you list the paternal genotypes across the top and the maternal genotypes down the left side. Now multiply together to represent each child.

1/4 of the children are a//a and would exhibit the alpha disease. 2/4 children are "carriers" like the parents. 1/4 of the children are A//A and have two "wild type" alleles.

	A a	
A	A A	A a
a	A a	a a

Many processes are organized in pathways with multiple step; each step requires at least one gene; absence of any gene in the pathway can give a phenotype, for example

step 1 => step 2 => step 3 => step 4 => Color

Loss of 1, 2, 3, or 4 will be colorless. Many diseases and human phenotypes are complex because many genes are involved. Good examples include heart disease, cancer risk, height, red hair, old-age dementia.

12. Each distinct allele will have a different order of DNA bases, A G C T. Therefore alleles can be distinguished by DNA sequencing. A mutation is a change in one or more bases in the gene DNA that confers a phenotype. This is the basis for
    a. pre-natal diagnosis of many "disease states"
    b. associating specific alleles with increased risk for disease
    c. forensic identification
    d. determination of the exact strain of a virus or other pathogen
13. Allelic diversity is important for the survival of any species. Some alleles that are "deleterious" in one environment (sickle cell anemia in the U.S.) can be favorable in another environment (malaria-infested countries). Because the environment changes over time and is "unpredictable" there is no "best" genotype. Habitat preservation is guided by the desire to maintain a large enough population of a species to ensure its survival--another way of saying this is to maintain a "gene pool of allele diversity" over time to allow the species to survive in a fluctuating environment.

14. Natural mutations are common, and many have few consequences. Each person is a carrier of ~100 very deleterious alleles. This is our genetic load. It is the scientific rationale against mating with close relatives (you are likely to share the same deleterious mutant alleles inherited from the same progenitor).
15. Some deleterious alleles are common in particular populations. If 1% of the children born in a particular ethnic group have cystic fibrosis, how common is the allele in the adults of that population?

    CF is recessive (a loss of function disease)

    Equation $p^2 + 2pq + q^2 = 1$

    if q is the allele state for CF disease and p is wild type

    $q^2 = 0.01$ therefore "q" is 0.1 or 10%

    10% of the adults in that population are "carriers" wild type//CF

    Genetic tests are "imperfect"

    False negative--you are a carrier but your test results were negative

    False positive--you are positive in the test used but an "unknown" factor overcomes the detected defect.

A good test is considered to have less than 5% false positive and less than 5% false negative. Is this reassuring? Imagine a scenario of fetal testing for Tay Sachs (a lethal disease state in which babies die within 5 years) disease alleles. Many parents who are Tay Sachs carriers choose to abort a fetus that contains two "bad" alleles. If the test itself is flawed, 5% of the time a healthy fetus will be terminated. Similarly, 5% of the time a Tay Sachs sufferer is born.

Genes are composed of hundreds to hundreds of thousands of DNA bases. Mutations can occur at any base.

Often a test is designed to detect the most common "disease allele" (such as a mutation of a G to an A at position 100 in the gene) by restricting the test to a specific region in a gene. Disease-causing alleles at other positions may not be examined in the test;

this is the most common reason for a "false negative" result. To make tests more robust, human geneticists would like to sequence the "disease" alleles from many individuals with specific diseases and families with carriers; in this way tests can be designed that detect more of the variations in the gene associated with disease. How can privacy be guaranteed? Under what circumstances would you volunteer for your DNA? If your parents, siblings or other relatives learned about a genetic disease state they carried, should they share the news with you?

We all want our children to be better off than we are. We hope they will be better educated, healthier, financially more secure. 5-10% of children inherit and identifiable genetic defect. Some more serious than others i.e. asthma to sickle cell anemia. Till now medicine has largely been about treating the symptoms but now the hope or promise of Biotechnology is that embryos can be prescreened for disease and even diseases that have childhood or adult onset can be treated.

First we will look at what genetic changes give rise to genetic diseases, and what are some examples of these diseases. Second we will look at how biotechnology might treat various genetic diseases.

Genetic mutations can take the form of the loss or addition of a single of a few bases or large loss or rearrangement of chromosome pieces or abnormal numbers of whole chromosomes.

FORENSIC APPLICATIONS OF BIOTECHNOLOGY

The Case of the Frantic Freshman

You and a friend are investigated as the possible perpetrators of a crime, based on a close match between your physical descriptions and the reports of eyewitnesses. The crime takes place in Branner and was very likely committed by a Stanford student. No strangers were seen in the dorm. The perpetrators must have worn gloves, because there are no fingerprints at the crime scene. There are hairs from the perpetrators, pulled from their heads by a low shelf above the crime scene. As one of the police remarked "I can hardly believe 2 people managed to squeeze through that space." The hair colors and general texture match you and your friend. You are arrested and charged with the crime.

The police offer you the chance to volunteer additional hair and blood samples for

a. Microscopic examination and comparison to the crime scene hair
b. Extraction of DNA from your hair to match to the DNA in the hair follicle cells present on a hair shaft that is pulled from the skin.

Because you are well versed in the power of DNA technology, you are sure that a valid DNA analysis will clear you and your friend. After all, you did not commit the crime, and it is only the "mistaken" eyewitnesses who judged you on the basis of gross physical characteristics (approximate height and weight, hair color, approximate age, body build, clothing) to place you at the scene of the crime. You argue that even if the test is based on just 10 genes, each with two alternative alleles + and-, and each allele is present in a 50% frequency, the "odds" of a match are 1/2 to the 10th power = or really low.

Thus, with 1600 freshmen, about 6000 undergraduates and 7000 graduate students on campus, the test is bound to demonstrate that neither of you were involved. Your friend is worried though, because as a statistics freak s/he is concerned because a true proof of innocence may be very difficult to obtain from the DNA samples. S/he becomes acutely worried when your lawyer tells you that the "classic Harvard medical student group" is the comparison population (1000 graduates of the Harvard Medical School, 1950-1955).

Polymorphism at 10 markers was assessed in this group. A surprisingly large number of alleles were present in 1% or fewer of the Harvard doctors. Thus, certain combinations of markers (1/1000)(1/100)(1/100)(1/100) should be exceedingly rare, present in just one or a few people on earth today.

You are allowed to request a few books. The dog-eared (from extensive study) copy of your Essential Biology textbook arrives and you excitedly show your friend how a few PCR tests will demonstrate that both of you are unrelated to the perpetrators. Your friend overwhelmed by your persuasive logic--relents, and both of you provide hair samples and settle in for a long weekend in jail.

Three days later the test results are back. First, you learn that the test was stopped at only 3 gene tests. Second you learn that your friend is "proven not involved beyond a reasonable doubt" because there were 0/3 matches to the crime scene sample.

You are likely guilty because the match was 3/3 to one crime scene sample; the markers employed were all present at 1% frequency in the comparison human population indicating a 1 in 1 million chance of individuals having all 3 markers (1/100 X 1/100 X 1/100). Thus one person is cleared because they matched none of the markers and one person is "guilty" because they matched all three tested. You scream, "More tests are needed!" awakening the entire jail.

The police tell you "The people" want justice, not more tests. And, besides the tests are great--proving for sure that your friend wasn't there, just you and someone else. Confess now, and beg for mercy. You're just a freshman, who knows, the punishment may be light."

"But, I didn't do it, and I can prove the tests so far are based on a faulty assumption, " you plead. Some details are as follows:

1. Each person-except identical twins-has a unique genotype inherited from the "recombined" alleles in the sperm and egg that formed the new individual. Most alleles differ by single nucleotide substitutions, also called SNPs (single nucleotide polymorphisms). SNPs are often diagnostic for alleles associated with specific diseases. Each SNP associated with a specific gene.
2. Some regions of the genome undergo rapid changes, resulting in great diversity of allele types. The best-studied examples are simple sequence repeats; these sequences are found at multiple locations in the human genome

(multiple loci). The simple sequence repeants can amplify or shrink, resulting in length polymorphisms at many locations. These polymorphisms can be measured by the polymerase chain reaction resulting in a genomic "fingerprint" characteristic of each individual.

3. Some length polymorphisms are common-found in 50% of humans-while others are very rare-found in 1 person in 10,000.
4. In a forensic application, the first check is to ask if a DNA specimen at the crime scene matches the "victim." If not, the DNA sample from hair, blood, skin, or semen is presumed to be from a person who frequents the crime scene or from the perpetrator. People innocently associated with the crime scene may match samples found there, i.e. family members have undoubtedly left samples in their home. Matches to family members or others who have had access to the crime scene are used to pinpoint any "novel" DNA sample; if such a "novel" sample exists, it is usually assumed to be from the perpetrator.
5. Likely suspects are often asked to contribute a DNA sample for comparison to the "novel" sample at a crime scene. The odds of a match are calculated from the known frequency of a particular "polymorphism" in the population. For example, in California only 4% of the population has type AB negative blood. If this blood type were designated as the "perpetrator" type, then all suspects with O, A, B, or AB positive blood types would automatically be eliminated as suspects. Only those suspects with AB negative blood types would still be under suspicion. Of course, common blood types such as O positive (~50% of the population) are not very effective at ruling people out as suspects.
6. In modern applications, multiple DNA markers are used-typically length polymorphisms at simple sequence repeats. For each marker, there is a pre-existing frequency known in the population (just like with the AB negative blood type).

Polymorphism Scoring and Frequency

A not a	B not b	C	D	E	F	G	H	I	J	K
0.1	0.2	0.5	0.5	0.08	0.01	0.1	0.4	0.1	0.05	0.2

In this hypothetical example, locus "E" is found in 80% of the population, and locus "F" is found in just 1%. Because there is an independent chance of inheriting a specific length at locus A, at B, at C, etc. the probability scores get multiplied together to give the final proportion of the population who would be expected to be ABCDEFGHIJK rather than any combination with a lower case letter, i.e. aBCDEFGHIJK.

In this example the cumulative "odds" are 1.6 x 10-11 of a person having the ABCDEFGHIJK genotype. Given that there are only 6 x 109 people on the earth today, this evidence would seem to indicate that the suspect is the only person who could have committeed the crime.

7. What could be wrong with this conclusion?
   a. One or more of the tests is flawed, i.e. there is human or machine error in distinguishing "C" from "c". To account for this, the "error" associated with each measurement must be known or more commonly the sample is tested twice and only if both tests agree is the data included in the final calculation. Science can not guard against fabrication of test results-recently a forensic testing in Oklahoma was accused of finding the data that matched the "likely suspect" rather than finding the data and then using that to exclude suspects.
   b. The frequency in "the" population distorts the results to implicate all/many/some/several members of the same group. If the allele frequency for A is 1.0 in Navajos, then an A result is consistent with all Navajos, not a specific Navajo person.
   c. The origin and composition of the population is crucial to interpreting the results. Our Frantic Freshman was

implicated based on 3 tests and the friend was eliminated. Some markers are from the "Y" chromosome-women are always missing these markers and hence can be quickly eliminated as a class from the list of suspects. One interpretation of the story is that the friend is female and the Frantic Freshmen is a male who happens to carry "common" alleles for the 3 loci examined. Perhaps the 3 markers are each 0.5 in the entire American population: 0.5 x 0.5 x 0.5 = 0.125 or 1 in 8 males will be the same genotype as the Frantic Freshman. This is not a very definitive test.

d. The Frantic Freshmen was compared to a data set comprised of medical students at Harvard, a long time ago. In that era, nearly all medical students were white Christians and of northern European heritage. The occasional Jewish, Hispanic, Italian or Greek class member could have alleles found in no other member of the class. This likely explains why a number of alleles were present at 0.01 frequency-one person in the 100 member class was of this genotype.

e. The Frantic Freshman comparison group is not representative of current Stanford undergraduates. Any person unlike that comparison group in heritage might be implicated "by the odds" and still be innocent.

8. Successful application of DNA forensics requires demonstration that the genotype is not just associated with heritage but is a true identifier of an individual. With 10 markers, each present at 0.5, there will be 1 person in 1024 that matches. Ancillary information can be used to firm up suspect identification. For example, pet hairs (pet DNA) associated with the victim can be sought in the suspect's car, clothing or home. Again, individual dogs and cats have a "genetic fingerprint." If a particular pet hair is likely to be found in 1/1024 animals and the suspect is a match at 1/1024, the odds can be multiplied to yield a probability of 1/1,048,576 that both the suspect's DNA at the crime scene and a particular pet DNA in the suspects

possession would be consistent with the crime. Similarly, victim DNA should not be associated with a suspect who is a stranger.

The ability to determine genotype raises many questions about individual freedom and privacy.

Agrobacterium tumefaciens, a soil bacterium, can transfer pieces of itsTi plasmid DNA to plant cells. This bacterial DNA is integrated into the plant chromosomes, and it is expressed.

In nature, the transferred DNA (T-DNA) encodes enzymes that cause the infected plant cells to proliferate producing crown gall disease; this is a form of plant cancer. The cancerous plant cells secrete opines (amino acid derivatives) that are aphrodisiacs and food for Agrobacteria. The purpose of the "natural transformation" appears to be a means for bacteria to get plants to use their amino acids to produce a specialized diet for the bacteria.

Ti = tumor inducing plasmid

T-DNA = DNA transferred to plant cells

The Ti plasmid was genetically engineered beginning in the early 1980s to deliver recombinant DNA to plant cells. The "opine genes" were eliminated from the Ti plasmid so that the recipient plant cells would exhibit normal growth. Initially, plant transformation was performed on leaf disks-the disks were cultured on agar plates to coax the plant tissue to regenerate into a whole plant. Today, whole plants are simply dipped into a solution containing Agrobacterium to introduce the DNA into cells within the plant. The dipping is often done in a partial vacuum to cause the fluid to infiltrate into the plant cells (like sucking on a straw to pull in fluid, you create a partial vacuum in your mouth). Transformed cells in the floral buds can become pollen (carrier of sperm) or eggs, insuring transmission to the next generation.

Transgenic plant = plant containing a piece of DNA introduced by genetic engineering; the DNA could be from the plant itself!!! More commonly, the introduced DNA confers a novel property. Transgene = the genetically engineered gene.

Some plant species are resistant to Agrobacterium; many of these species can be transformed using the gene gun. Gold particles are coated with plasmid DNA containing the transgene; these particles are accelerated and "shot" through the plant cell wall. The DNA diffuses off the gold beads, and some is incorporated into the plant chromosomes. This method is more difficult and less frequently successful than nature's genetic engineer.

Initially, genetically engineered plants were resistant to specific chemical herbicides. This was "the strategy" of herbicide manufacturers to sell the special seeds and more of their herbicide. Round-up resistant soybeans and corn are now grown widely in the U. S. These herbicide resistant crops have been very controversial-and the source of much of the criticism of transgenic crops.

Modern goals of genetic engineermg of plants include:

a. improving plants for human or animal nutrition, i.e. with increased vitamin content

    i. plants synthesize Vitamin E (alpha-tocopherol) and are the primary dietary source, other than eating fungi or vitamin pills. Humans need Vitamin E to protect against oxygen damage in the cells; Vitamin E is currently thought to help protect against heart disease and cancer by reducing the "free radicals (active oxygen derivatives)" in cells

    ii. in plants, Vitamin E and related compounds (other tocopherols) are potent anti-oxidants that protect plant cellular contstituents from reactive oxygen generated as a by-product of photosynthesis. Typically only 1-10% of the "Vitamin E-like" compounds are the form most useful to humans

    iii. the "related compounds" are not useful to humans or other mammals because these compounds are not absorbed from the food in the diet

    iv. Arabidopsis and other plants typically accumulate a tocopherol "one step" before Vitamin E; by making transgenic plants that "cver-express" the last enzyme, most of the tocopherol is converted to authentic Vitamin E (up to 90%)

v. transgenic plants are otherwise normal. Oil made from the transgenic seeds supplies a high daily dose of 400 International Units (50 is the minimum daily requirement) in just one tablespoon of oil. The non-transgenic "wild type" plants synthesize so little Vitamin E that one cup of oil would be required

vi. some day your "healthy" salad may contain lettuce with enhanced levels of Vitamin C, carrots with extra carotenes, and salad dressing supplying your Vitamin E for the day

b. making plants that can manufacture a rare compound (such as an antibody used to treat patients) or that can synthesize a product such as plastic, replacing chemical synthesis

c. having plants deliver a vaccination dose in the food

i. transgenic plants synthesize part of the hepatitis virus or part of the bacterium that causes diarrhea

ii. in raw food, the transgene-encoded novel protein is an antigen that your body recognizes as foreign; the antigen elicits antibody production, protecting you against the disease-causing organism; the antigen is destroyed by cooking at high heat, ruling out putting the transgene into potatoes or other foods that are cooked prior to consumption

iii. the "plan" is to make transgenic bananas each expressing a different antigen; by eating a few bananas people, particularly children, could be immunized without expensive needles or sterile vaccines; bananas store well, the fruit is sterile inside, and the taste is palatable to nearly everyone

d. designer plants just for entertainment-plants with the luciferase gene from fireflies to glow in the dark

Modern genetic engineering strategies must contend with consumer and government suspicions about transgenic crops.

The legacy of the initial herbicide-resistant products is that many people distrust transgenic crop technology.

You are in charge of developing policies to regulate transgenic foods, to decide which applications can enter the marketplace immediately and which require further tests. To implement your policy decisions, you will need a "education" campaign such as the USDA "food pyramid" to guide diet choices or the anti-smoking ads.

Before thinking of the SLOGAN and SONG for a transgenic technology ad campaign, you need an assessment of the perceived possible dangers, how much risk is associated with each possible danger, and whether there is a simple remedy (for example labeling). In preparation for class discussion determine your risk assessment. Is there an appropriate remedy or solution? How will risks without remedies be explained to the public? Could additional research provide a remedy? Should the technology be delayed to await more data?

Some thoughts on bio-terrorism:

- Chemical and biological terrorism has long been with us. In the United States as long back as 1754 when smallpox infected blankets were distributed among the American Indians.
- In 1984 750 people were poisoned in Oregon with salmonella bacteria.
- Sarin gas which blocks acetylcholinesterase killed 12 in 1995 in Japan.
- Larry Harris a member of a white supremacist group ordered Yersinia pestis from the ATCC in 1995. (Y. pestis is the agent of the bubonic plague)
- Ebola virus kills 90% of those infected. There is no treatment, no cure. The mode of transmission is unclear.
- Kathleen Bailey, former assistant director of the U.S, Arms Control and Disarmament Agency is convinced that for $10,000 and a 15X15 room a major biological arsenal could be built with no special gear.

- It is suspected that at least 20 countries have violated the 1972 Biological Weapons Convention-a treaty that prohibits the development or possession of biological weapons.
- Laws prohibiting poison go back to the Greeks and Romans who condemned the use of poison in war as a violation of ius gentium-the law of nations. The Manu Law of India in 500BC also outlawed poison.
- Agricultural biowarfare is a viable concern. In the U.S. there have been programs to develop plant pathogens of drug crops as biocontrol agents. Iraq was developing wheat smut as a weapon for use against Iran.
- In 1996 the United Nations estimated that $80 billion would be necessary to stockpile indicated numbers of gas masks, antibiotics, vaccines and for civilian defense measures.
- Many feel the use of chemical weapons by Iraq against Iran in 1995, of anthrax and botulinum toxin, with no political consequences has lowered the political threshold for the use of bio terrorism. It seems clear that whereas biotechnology could be employed to make additional lethal microbes, there is a sufficient supply in nature already.

Facts about Anthrax

Caused by the spore forming, gram positive, aerobic bacterium, Bacillus anthracis. Of the 211 cases in the U.S. from 1955-69 in humans 33 were associated with wool and 106 with imported goat hair. In 1877, Robert Koch used anthrax to prove that a bacterium was responsible for infection in animals. His experiments established Koch's postulates. It was also with anthrax that Pasteur and Jobber, in1880s, first noted the protective properties of Pseudomonas aeruginosa, the first hint at antibiotics. (Note Fleming didn't discover Penicillin till 1929)

Before 1954, death in animals due to anthrax was simply thought to be due to the sheer numbers in the blood stream, that literally the capillaries were blocked. The anthrax toxin is however still not completely understood. Three different factors carried on a plasmid contribute to the toxicity. Death occurs due to eliciting an immune response where the macrophages are induced into a mass release of inflammatory hormones resulting in increased vascular permeability, respiratory failure and cardiac failure. One reason antibiotics don't stop anthrax death is because if toxin is present in lethal amounts by the time antibiotics are given, then they are of no use. A polysaccharide capsule serves the bacterium by aiding in infection. Vaccines for anthrax have resulted in severe side effects based on a study done with 2.5million members of the military. The vaccine consists of a series of six doses with yearly boosters. How is biotechnology playing a positive role?

1. Trace source of the strain. By determining the DNA sequence of the terrorist anthrax and comparing them to the libraries of known anthrax, hopefully it will be easy to determine how related the strains are and where the terrorist strain originated. Thousand of different anthrax strains are in the databases in the Los Alamos National Laboratory
2. DNA hybridization with specific probes could quickly identify the strain.
3. Use pure antibodies to quickly test and determine where anthrax antigens are present.

How might biotechnology playing a negative role? A few might be

1. Introduction of antibiotic resistance to strains such as B. anthracis or Y. pesti.
2. Glean from the sequence databases information to make a designer lethal bacterium.
3. Random mass selection for toxicity or resistance.

Prokaryotes: Life as we know it depends on bacteria, to provide vitamins, protect us from disease and decompose organic matter. Bacteria can be found most anywhere. We have already talked about how the enzymes from the hot springs bacteria Thermus aquaticus changed the way we do sequencing and allowed for the technique of PCR. Bacteria come in several shapes: cocci, bacilli and spiral are the most common. Today we will discuss two bacilli.

BACILLUS THURINGIENSIS

First discovered in 1911 as a pathogen of flour moths in Germany it was first used as a commercial insecticide in France in 1938, and later in the US in the 1950s.

Since this time more effective strains have been found and marked worldwide as an insecticidal bacterium for many important plant pests. Particularly toxic to caterpillars and also mosquito larvae and blackflies. Bt "products" represent about 1% of the total agrochemical market worldwide.

This Bt product consist of dried spores and toxin crystals that are applied to leaves and other environments where the insect larvae feed. Only a relatively few bacteria can form endospores. Bt can as well as Bacillus anthracis, which we've been hearing so much about. These endospores can remain dormant for years, even centuries, and are very resistant to the environment.

THE PROTEIN

Before we talk about the cloning of the Bt gene, let's understand the protein. The crystallize toxin is a large insoluble protein which is safe to humans, animals and most insects. Once solublized in basic conditions of above pH 9.5 (the conditions found in the mid-gut of a butterfly or moth) the protoxin is cleaved by proteases in the gut to the smaller active toxin.

The toxin binds to the epithelial cells in the gut, creates pores in the cell membranes resulting in an equilibration of the cellular ions. The cells lyse, blood pours into the cells, lowering the pH, allowing the bacteria spores to germinate and grow throughout the host. The larvae stop eating immediately and die several days later.

Note the Bt toxin is very closely related to diphtheria toxin. Of the crystal toxins (cry toxins) there are over 50 different ones that fall into 15 different groups. Bt also can carry a second toxin known at the cytolytic toxins. Typically these toxins are carried on plasmids and can be moved from strain to strain via conjugation. Transposons within the bacteria can also move the DNA to create a more diverse family of toxins. Resistance to the toxin is usually seen associated with a change in the binding of an essential gut enzyme, for example aminopeptidase-N.

Plants genetically engineered with the Bt gene:

Since 1996, the Ti plasmid from Agrobacterium tumefaciens was used as a vector to introduce Bt toxin genes into many plants; corn, potato, cotton, canola and soybean. This translates into hundreds of millions of dollars worth of Bt stuff!

Let's look at some of the numbers:

109 million acre of farmland worldwide are planted with GM crops-Primarily corn, potato, cotton, canola and soybean (almost twice the area of the U.K.) (68% in the U.S. alone)

Every year 971 million pounds of pesticides are used to kill insects, weeds and fungi in the U.S.

Bt cotton according to the EPA has resulted in a 21% reduction in insecticide use on this crop. (Bt corn and potatoes report less of a pesticide reduction partly because they use fewer pesticides to begin with)

Concerns: How do Bt crops affect "non target " organisms? Bugs birds and worms? Will GM crops pollinate nearby plants?

Studies have shown that green lacewing caterpillars were more likely to die after eating European corn-borer caterpillars that had fed on Bt corn.

John Losey at Cornell University reported that milkweed leaves dusted with Bt corn pollen to monarch butterfly larvae died. Although overall this risk is thought to be small. (Less than 150 pollen grains per square centimeter of leaf). One suggestion is to have a buffer zone of normal corn will "protect milkweed" found outside the fields.

Could Bt cause allergic response in people? StarLink Bt (Cry9C) corn was found in trace amounts in Kraft taco shells. StarLink had only been approved for animal feed because it was 50-100 times more potent than other Bt varieties and there was concern that is could act as a human antigen. 2.5 million boxes of tacos were recalled, and since this initial discovery StarLink has been found in many of the brands we know and love, Kellogg, Archer Daniels Midland and Tyson to name a few. 8% of all US children suffer from food allergies and this concern is increased by the possible correlation between food allergies and asthma.

Unrelated to Bt, but related to herbicide resistant plants there is concern that "superweeds" will arise. Could pollen from herbicide resistant plants pollinate weeds that now would be resistant to known herbicides?

In a ten-year study no superweeds have been found, however GM canola plants in Canada have invaded some wheat fields as if they were weeds. Concern for viral resistance may flow into the wild populations resulting in such things as wild oats over running cultivated plants. One solution is to insure GM plants are not grown anywhere where close plant relatives might live. This practice is already being done in south Florida and Hawaii for cotton.

Bt crops will someday fail. Concern centers on the fact that continual exposure to the toxins creates a strong selective pressure for development of resistance to the toxins.

To address this concern the EPA requires non-BT crops, Bt refuges, to be grown on some part of the Bt grown farmland to help dilute out a possible resistant trait. (This year the requirement is 20% for corn and 50% conventional cotton.) So far there have been no reports of insect resistance to Bt. The bacterium Photorhabdus luminescens has a similar and possibly more potent toxins which can also be moved into plants much in the way Bt has.

The EPA is currently determining whether Bt corn and cotton should continue to be grown in the U.S. In March of 2001 the Scientific Advisory Panel cited weaknesses in the EPA's current Bt program:

a. The agency has yet to develop an adequate ecological risk assessment process.

b. The approach for evaluating human health risk of Bt crops needs to be broadened.
c. The Bt corn impact on monarchs remains unanswered.
d. The EPA needs to ensure that Bt crops are grown under stringent resistance management plans to delay evolution of pest resistance to Bt toxins.

The EPA decision on Bt corn and cotton is due out any day.

Concern centers on the fact that continual exposure to the toxins creates a strong selective pressure for development of resistance to the toxins. What sort of crop management strategies might minimize this risk?

Whereas most GM crops will not be of environmental concern, it remains to be seen which one are. How do you feel about salt tolerant rice?

FACTS ABOUT SMALLPOX

Smallpox is a viral disease unique to humans. It is a double stranded DNA enveloped virus. It is spread from person to person by inhalation of air droplets or aerosols.

Twelve to 14 days after infection the patient usually has a fever and aching pains much like the flu. A couple of days later the rash develops first on the face then on the extremities. Death usually occurs during the second week killing 25-30% of those unvaccinated.

There is no treatment, only isolation of the patient. Vaccination within 2-3 days of exposure affords almost complete protection against tne disease. Vaccination 4-5 days after exposure seems to give some protection from the virus.

Why not just vaccinate the entire population? Aside from the cost, the vaccination has several risks:

- 3/million will result in encephalitis and 40% of these cases will be fatal, others are left with permanent nerve damage.
- A second risk is that the vaccinia vaccine may continue to spread and grow, know as Progressive vaccinia disease,

this disease is common in patients who are immuno suppressed, for example AIDs patients and patients undergoing chemotherapy.

Currently in the US there is vaccine stocks for 6-7 million people. It will require 36 months to increase this stock. In 1972 the US eliminated the requirement of a smallpox vaccination for children entering school. The last cases of smallpox had occurred 23 years before in the US.

Some Recent Incidences of Smallpox

1970-In Germany, an electrician was returning from Pakistan. He was admitted to the hospital where at first there was concern he had typhoid fever. 5 days later a diagnosis of smallpox was confirmed.

He was transported to one of Germanys isolation hospitals and 100,000 people were vaccinated. Hospital staff and patients were quarantined for four weeks. 19 people came down with smallpox. One was a person who had spent only 15 minutes in the hospital and had only briefly been 30 feet from the patient's room.

1972-In Yugoslavia, a pilgrim returning from Mecca became ill, many friends and relatives visited the patient. Two weeks later 11 of the visitors were ill with high fever and rash. These people were typically seeing different physicians and a conclusion of smallpox was not drawn. (This is one serious problem that most physicians have never seen smallpox!)

Four weeks later the correct diagnosis was made. By then 150 people were ill. At this point 20 million people were vaccinated. 10,000 people or more spent 2 weeks in isolation. Neighboring countries closed their borders. Nine weeks after the first person became ill the outbreak stopped. 175 patients has contracted smallpox 35 were dead.

It is guesstimated that 10-15% of the US population has some residual resistance to smallpox.

Can weird things happen when biotechnology changes a virus?? Yes

In Canberra, Australia the mice population is of severe concern. Two scientist, Ian Ramshaw and Ronald Jackson were working on a contraceptive vaccine for mice by altering the genes of the mousepox virus. Yet in February they published in J. Virology that they had inadvertently created an unusually virulent strain of mousepox.

Their approach was to take a benign strain of mousepox virus and add genes for proteins carried on the surface of mouse eggs. The concept was that cells infected by the virus would make egg proteins on the outer coat. In turn the female mice would produce antibodies against their own eggs.

To make the vaccine really effective Jackson and Ramshaw also put in the gene for interleukin-4 (IL-4) a protein that usually boost antibody production.

This backfired! The IL-4 gene also shut down the animal's immune system resulting in their susceptibility to the usually mild mousepox, and even those vaccinated against mouse pox were killed. Could the same thing be done with smallpox? Should this paper have been published?

Typically hybrids are much less virulent than original wild type strains.

And don't worry evolution is on our side. Nature is very efficient and doesn't like to keep genes around unless they are going to be used! For example a bacteria doesn't easily maintain resistance genes unless they are continually selected for.

Why is there no treatment for viral diseases? What are the three main classes of antibiotic and how do they act? Think about protein synthesis; cell wall synthesis and folic acid synthesis.

VIRAL REGULATION

You are now familiar several vectors used to clone DNA: plasmids, Ti plasmid, BACs, YACs, phage vectors and transposons. Of the nine different types of viruses two of the most promising viruses for gene therapy are the adenoviruses and the retroviruses.

Adenoviruses-are double stranded DNA non-enveloped viruses of medium size that are usually associated with respiratory infections. We have discussed these viruses as a mode for introducing a healthy cystic fibrosis gene. Adenoviruses do not integrate into the host DNA. Thus any descendants of the genetically engineered cells do not carry a functional cystic fibrosis gene. Once the genetically engineered cells die the functional cystic fibrosis gene is lost. As a result people with CF would need to be retreated every few months.

Retroviruses-Include all RNA tumor viruses and single stranded RNA viruses. These viruses cause leukemia and tumors in animals and the Lentivirus HIV cause AIDs. As and example of how a retrovirus works the AIDs virus the steps are:

a. HIV circulated through the blood and lymph system until one of its extracellular proteins (gp120) binds to host cells that bare the receptor CD4. This binding results in a fusion much like two soap bubbles coming together.
b. Inside the cell, enzymes in the cell strip off the viral protein coat. This "uncoating" process releases the RNA genome and the reverse transcriptase.
c. Reverse transcriptase uses the RNA as a template to make a double stranded (ds)-DNA strand.
d. The ds-DNA is transported into the nucleus where it is integrated into the chromosome of the cell, and is now known as the provirus.
e. Here the provirus can lie dormant or begin to produce new viral RNA and viral proteins.
f. Final assembly of the virus occurs at the cell membrane, where the new formed virus will bud out through the membrane. However the viral proteins (such as gp120) will be found on the outside of the membrane

Can you think of how antibodies or PCR might be used to test for HIV infection?

Can you see how a retrovirus might be a good candidate for gene therapy?

One problem is that retroviruses only infect dividing cells. Nerve cells are not dividing cell; so many diseases can not be treated with retroviruses.

A second problem is how to direct where the retrovirus inserts into the DNA. Until now there is very little control of insertional position. The regulation of random insertion can be a problem. Disruption of tumor suppressor genes in monkeys by retroviruses has been shown to cause cancer.

In conclusion, you can now see that we can sequence, clone, and move genes into cells but one of the last areas of complicated work is gene expression and gene regulation.

7

BIOLOGY IN HUMAN MEDICINE AND HEALTH

INTRODUCTION

Should pigs be pharmed for replacement kidneys? How do our cultural myths--the Frankenstein story and the Bionic Man--influence acceptance of replacement organs? Should animals be cloned to further the goals of agriculture or to produce products for human medicine? If replacement parts are in limited supply, who decides the recipients, and by what criteria?

The world of developmental biology changed dramatically in 1997 with the birth of Dolly the sheep. For more than 100 years embryologists had tried to induce adult cells to regenerate into embryos or had tried transplantation of adult or even embryonic nuclei into eggs. Many of these experiments were failures because so little was understood about the key stages of post-fertilization development, particularly in mammals. Mammals have internal fertilization and the embryo implants into the mother's tissues; these features made it difficult for scientists to observe how normal embryos developed.

Although studies on animals are usually "the model" for advances in human research and medicine, the opposite is true in "animal cloning" studies. Rapid advances in assisted reproduction technologies for humans--particularly in methods of harvesting viable eggs and of maintaining in vitro fertilized embryos in culture prior to transplantation into a surrogate mother-

-provided key insights into the rate-limiting steps of reproduction. For the past 30 years, research on human reproduction has provided a continuous stream of important technical improvements in mammalian reproduction. It may surprise you to learn how far research has advanced in culturing human body parts and in using human genes in animals.

SOME DEFINITIONS

Replacement parts from "humanized" Animals Transgenic animals can be generated that express "human" proteins in specific organs. There are already mice that produce "human" antibodies. What if pigs could grow kidneys that matched human kidneys? Such kidneys could be used for transplantation into humans with a similar "antigen" profile (like blood typing, all organs can be typed for a good match between a recipient and a donor).

Steps

1. Select an appropriate animal model species with similar organ size to humans.
2. Transform the animal with human genes that are expressed in the target organ to "humanize" the organ for transplantation.
3. Demonstrate that the animal is healthy and free of viruses that might cause disease in humans.
4. Not Yet Approved in the Us. Acquire permission for human testing via transplantation of the "humanized" organs. Typically, the initial volunteers for such experiments are very ill-the experimental protocol may be their last chance for a cure. Do you think that such ill people freely ascribe to "informed consent" of the procedures and implications of experimental therapies?
5. Screen human patients for a good donor match to the pharm animal humanized organ.
6. Sacrifice the animal (or at least remove some of its organs) and transplant them into the human volunteers.
7. Possible in the Future if a particular animal was a highly successful organ donor, this animal would be selectively

bred or possibly even cloned to produce near-identical or identical offspring as a continual source of new "humanized" organs.

A somewhat science fiction version of the replacement parts scenario would be growing "humanoid" cell masses in test tubes as a reservoir of parts. This can already be done for some tissues such as skin. Burn victims may have skin patches removed from unharmed areas of their body; these patches can be grown into large sheets of replacement skin in the lab and then grafted back onto the body.

A baby's placenta could be recovered and grown into a reserve of some body parts, particularly blood cell types. Placentas are frozen at many hospitals, and kept for about two years to treat childhood blood diseases such as leukemias. Would growing human hearts, kidneys, lungs, livers, skin, etc. in a laboratory be an ethical problem? Should people be encouraged to donate organs for laboratory propagation?

Should we be producing many kinds of pharm animals to assist humans quest for health? Would vegetarians participate in therapy based on animals? Would you?

Somatic Gene Therapy Introducing genes into the somatic body of a human patient to correct a phenotype by supplying new genetic information, i.e. "mutant" blondes could acquire better hair pigmentation. Juvenile onset diabetics could receive better insulin genes, etc. The somatic replacement genes are NOT transmitted to the next generation.

Steps

1. Identify and clone a functional allele that can cure a particular genetic disease.
2. Construct vectors that contain all of the information important for the correct quantitative and qualitative aspects of gene expression. This can be a long procedure of verifying what parts of the chromosome are relevant.
3. Recruit human volunteers to receive the gene therapy after receiving permission to conduct experiments on

human subjects. The patients must give informed consent to the experimental procedure.

4. Transform their target tissue-this could involve injecting DNA into a particular organ and hoping that cells take up the DNA, integrate it into chromosomes, and express it. Alternatively, a human virus could be used to deliver the DNA to its target cells.
5. Monitor the patient for amelioration of symptoms and for expression of the new gene. It is important to show that there is a correlation between the treatment and the symptoms, because there will not be a placebo "control" group for such a risky experimental procedure.

DISCUSSION ON SOMATIC THERAPY

1. Should somatic gene therapy be performed?
   - Safety concerns: Will the introduced gene have side effects?
   - Can the gene expression (or the genes themselves) be targeted to the correct cells in the body?
2. For what phenotypes should we invest public research funds?
   - Severity of disease?
   - Likelihood of amelioration by somatic therapy?
   - Should diseases with an adequate current therapy (diabetes) be explored if they impact a large number of people?
3. Should the private sector be allowed/encouraged/prevented from developing therapies for phenotypes that are not "diseases" but for which there is substantial citizen interest? Examples such as hair color, increasing height, etc. come to mind.
4. Would you participate in a somatic gene therapy trial?

DISCUSSION OF REPLACEMENT PARTS

Is research on human replacement parts ethical when there are many people who do not receive even basic medical attention?

Are replacement parts for diseases with a strong behavioral component (smoking-induced lung cancer, adult onset diabetes) appropriate targets for research and investment of health dollars? Who decides who gets the replacement parts?

What fraction of human function or parts can be replaced in an individual and that individual still retains their own identity?

Would you choose to have "broken parts" replaced to live a longer life? Are there hidden costs to society of having such bionic individuals?

How do replacement parts differ from cloning?

HUMAN GERMINAL GENE THERAPY AND HUMAN CLONING

Germinal Gene Therapy and Human Cloning: In germinal gene therapy the genotype of a gamete (sperm or egg) could be corrected. A functional copy of a defective gene could correct a lethal condition such as Tay-Sachs disease.

The "replacement gene" would be transmitted to the next generation, following Mendel's Laws. You may want to work out on a piece of paper what the inheritance would be for an individual who has two defective copies of Tay Sachs at one locus and a single functional gene copy on a different chromosome. What % of the sperm or eggs produced by this individual will contain the functional copy? What happens to the defective copies?

Steps

1. Identify and clone the relevant allele, as described in somatic therapy.
2. Introduce this gene into an egg or sperm cell.
3. Conduct in vitro fertilization to insure the union of the "transformed" gamete with the opposite sex to produce a transgenic zygote. The resulting person WILL transmit the trait to the next generation.

2B. Remove a blastocyst from a developing human embryo and introduce DNA into the embryo cell.

3B. Coax the blastocyst to regenerate into a complete embryo, resulting in a clone of transformed cells. The resulting person WILL transmit the trait to the next generation. Or

3C. Put the transformed blastocyst back into the original embryo creating a mixed population of transformed and non-transformed cells. The resulting person will be a chimera of "disease" and "corrected" cells. If the proportion is favorable, they will likely be cured and MAY transmit the trait to the next generation.

Issues with Germinal Therapies

There are many pros and cons to germinal therapies. A primary motivator for all human in vitro fertilization work is the desire of particular people to have their own biological children (as opposed to having no children or adopting unrelated children). If you are "pro-life" you might consider that many people choose to terminate pregnancies when a fetus carries a lethal disease; a germinal gene therapy approach could allow proactive addition of the "missing" gene in both homozygous recessive (will die by age 2) individuals and supply an extra copy of the gene to the heterozygous carriers, reducing the chances that in the next generation they would produce gametes lacking this essential gene.

It is routine now after in vitro fertilization in animals to separate the blastocyst into several cell masses each of which can form a viable embryo. This is biological cloning of an individual. We also think of cloning using a somatic nucleus of an adult put into an oocyte. This is the procedure that produced Dolly the sheep. This procedure might allow families with segregating lethal genes to produce a heterozygous child from a nucleus of one of the two, healthy (heterozygous) parents with certainty. A father and son could be the same genotype (or mother and daughter), but would certainly have different life experiences.

Discussion of Human Germinal Gene Therapy

Biological Cloning. Why are you for or against it for animals? What if the animal is transgenic and making a product important in human medicine? Should germinal therapy on human embryos under any circumstance?

- Are there families that should be counseled that there are no "natural" reproductive alternatives for them? This is a statement about the extinction of a lineage (biologically).
- Should any research on this topic be supported by public funds?
- Should private companies be allowed to do research on this?

Should "twinning" of human embryos be allowed?

Should cloning of adult nuclei into enucleated human embryos be allowed?

Would you consider any form of germinal gene therapy or cloning?

Tissue engineering was so much simpler back in the days when it was so much harder. At least, from Alan Russell's perspective, it was simpler in ethical and moral terms. No one ever lost much sleep, for instance, over using foreskins to make artificial skin for burn patients, explained Russell, executive director of the Pittsburgh Tissue Engineering Initiative.

Today, ambitions have ballooned, as biotechnologists work on ways to treat strokes with neuron transplants, use polymer coatings to keep coronary arteries from narrowing and eventually grow replacement livers, muscles and other tissues. But one of the hottest topics in tissue engineering--embryonic stem cells--also is its most troubling.

Scientists believe almost any type of cell in the body could be produced from these embryonic stem cells, perhaps leading to cures for such diseases as diabetes, Parkinson's and muscular dystrophy. But the only way to obtain those cells is to destroy human embryos, a fact that has landed tissue engineering in the midst of the abortion controversy.

"There's a lot of 'Sophie's Choices' in this," said Russell, a biochemist and molecular biologist who heads the University of

Pittsburgh's chemical engineering department. Stem cells will be on the minds of 400 to 500 tissue engineering researchers expected to attend the three-day Engineering Tissue Growth international conference that begins tomorrow at the Marriott City Center. Several researchers will be presenting papers regarding stem cells--the adult-derived type, not the embryonic type--and how they might be used for such tasks as repairing the central nervous system.

Embryonic stem cell research, now performed by only a handful of researchers, could soon burgeon, depending on the outcome of a debate within the Bush administration and the Congress.

The National Institutes of Health last year issued guidelines for funding research on embryonic stem cells. The research would be performed using left-over embryos produced by in-vitro fertilization clinics that would otherwise be discarded.

Many abortion foes, contending that the unimplanted embryos are every bit as human as a fetus, have opposed the NIH plans. President Bush has expressed his personal opposition to this type of research and has asked Health and Human Services Secretary Tommy Thompson--an abortion foe but, as Wisconsin governor, a supporter of embryonic stem cell research--to review the guidelines.

Biotechnology experts worry that without such research, U.S. biotech companies might fall behind those in other countries, notably Britain, that allow embryonic stem cell experiments. Last week, a committee of the Canadian Institutes for Health Research proposed allowing embryonic cell research.

At the heart of the issue, said Michael Brannigan, executive director of the Center for Ethics Studies at LaRoche College, is whether an embryo has "moral status"--whether it has such basic human rights as the right to exist.

"The issue of moral status is one that philosophers have been debating for decades," Brannigan said. For those who believe embryos have moral status, it isn't possible to justify destroying an embryo to obtain its stem cells. Abortion foes, moreover, fear that allowing experimentation on embryos creates a "slippery slope" that could be used to justify abortions as well.

"One can't discredit that line of thinking," Brannigan acknowledged, though he personally is not convinced of an embryo's moral status. Proponents of the research maintain that the embryo doesn't have moral status, he added, and that manipulating or destroying embryos is justified, particularly if it is done to achieve a greater good.

Russell is personally conflicted. An embryo is human life, he said, emphasizing that he was not speaking on behalf of the tissue engineering group he heads. "It's not just a clump of cells; that clump of cells is a life." Yet he finds it difficult to conceive of even adult-derived stem cell research being able to advance without embryonic stem cell research.

Adult-derived stem cells, such as the stem cells in bone marrow that give rise to all of the blood cells, originally were thought to have much less potential than embryonic stem cells.

The adult cells normally produce only a narrow range of cells--blood cells, or muscle cells, or brain cells, depending on the type of stem cell. But experiments at Pitt and other research centers in recent years now suggest that adult-derived stem cells can be reprogrammed, so that a blood stem cell can give rise to, say, a liver cell or a brain cell.

Using adult-derived stem cells in this way would allow researchers to sidestep the moral minefield of embryos. "But the reality is none of us really know the potential [of adult-derived stem cells] yet," Russell said. For the foreseeable future, parallel experiments with both adult-derived and embryonic stem cells will be needed until scientists understand the capabilities of each type.

"You're going to have to find a way to find and collect embryonic cells if the work with adult-derived cells is going to make sense," Russell maintained. He admitted he doesn't know how that can be done yet.

Moreover, the federal government needs to develop a national tissue engineering policy to make sure these questions are addressed properly, he said. He fears that, if the NIH goes ahead with embryonic stem cell research, adult-derived cell work will languish as scientists gravitate to what appears to be the more promising cell type. Priorities should be set to maintain funding for both types of stem cells, he said.

Fusion of a haploid sperm and a haploid egg at fertilization produces a diploid zygote. The mammalian zygote is protected inside a covering the zona pellucida (produced by the egg) and is floating "free" inside the mother's reproductive tract.

Within four days, the single cell zygote divides to produce two cells; these two divide to produce four, and four divide to produce 8. Each of these 8 cells (blastocysts or blastomeres) is totipotent-this means that each individual cells has the ability to form a human. At the 8-cell stage, one or more blastocysts can separate from the loosely formed group; often this results in identical twins. Identical twins are the product of the same fertilization event.

After in vitro fertilization, a single blastocyst can be removed from a 3 or 4 day embryo for genetic testing: sex determination, disease diagnosis by PCR, screen for abnormal chromosomes. Many parents choose to destroy abnormal embryos at this point. When an embryo is demonstrated to be "normal" in all of the tests, it can be divided (artificially twinned) prior to implantation into the mother, or for frozen storage of normal embryos for future use.

Excess "normal" embryos are often available and have often been donated to scientific research rather than being stored or destroyed. These "extra" embryos are the source of most embryonic stem cell lines for scientific research. In a very few instances, private companies sponsored in vitro fertilization for the sole purpose of obtaining embryo cells for research.

At the next stage ~16-32 cells, some cells move to the outside to form a layer (trophoblast, which will be the embryos contribution to the placenta) around the internal cells (inner cell mass => future fetus). After this step, only the inner cell mass cells are totipotent.

The inner cell mass cells organize into a two-layered ball of cells: the outer layer (ectoderm) and the inner layer. At this stage, a portion of the ectoderm is already "fated" to adopt epidermal fates (skin, lining of the mouth and the rest of the gut, lining of the lungs, etc.). Inner cells are "fated" to produce the inner organs of the body. Cell fate means than barring intervention, the cells will naturally become a specific part of the body. As cells acquire a fate, they typically lose the ability to form other parts of the body; the process requires several steps. This restriction in cell totipotency means than cells from older embryos, from the placenta or from adult humans are unable to reform all of the cell types of the body.

Totipotency => pluripotency => multipotent => fully fated, specific cell type

Embryo cell => Bone marrow stem cells => blood cell precursor => lymphocyte

During gastrulation, a portion of the embryonic epidermis (outer layer) moves inside of the ball of embryo cells, in a process called invagination. Gastrulation establishes the three embryo layers: ectoderm (outer), mesodermal (middle), endodermal (inner most). Interactions between the layers, mediated by chemical exchanges between cells, results in the differentiation of specific organs and their constituent cell types. These inductive interactions depend on trophic factors (hormones and cell stimulating factors); in tissue culture scientists supply cells with such trophic factors to promote cell differentiation.

During neurulation, which starts while gastrulation is still proceeding, the rudiments of the brain and spinal cord are established, moving progressively from head to tail.

In quick succession, the major musculature of the body trunk and the limbs, the internal organs, and the organs of the head (eyes, ears, mouth, etc.) form. Neural crest cells migrate from the region above the spinal column throughout the body to form sensory hairs. You can see the pattern of this migration in tabby cats and zebras. During gastrulation the future germ cells migrate "outside the embryo" where they are not influenced by the inductive interactions that program cells to differentiate. Future germ cells later migrate to the genital ridge through the embryonic gut, where they stimulate the development of male or female-appropriate gonads. Prior to the arrival of the germ cells, the genital ridge is the same in males and females.

As organs mature, most retain some "stem cells" that will allow future growth and renewal of the organ. For example, there are stem cells below our "dead skin" that divide to produce new skin cells. In the bones of a fetus or growing child there are regions that can produce additional bone and cartilage cells to allow for extensive growth. These skin and bone stem cells cannot, however, produce diverse other cell types: they are multipotent stem cells, not totipotent.

By definition, when a stem cell divides (mitosis) one daughter cell remains as a stem cell and the other daughter begins to differentiate. In this manner, stem cells are "self perpetuating" and simultaneously produce the cells required to build organs and tissues.

When children or adults lose a key cell type, i.e. loss of the Islets cells during juvenile onset diabetes or dystrophy of major muscles, the phenotypes of the disease demonstrate that there are no longer stem cells (or enough stem cells) to replenish or repair the organ. Many scientists hypothesize that such disease states could be reversed or at least ameliorated by the additional of embryonic or adult stem cells that could proliferate to make new tissues of the appropriate type.

What limits application of stem cell therapy?

a. Adult stem cells have a limited ability to proliferate into the diverse types of cells needed to cure most diseases. One exception are the blood and lymph precursor cells in bone marrow; there are not many of these pluripotent stem cells, but each one can through cell division produce daughter cells that can replenish all of the white and red blood cells and all of the other immune cells of the body. After people receive whole body radiation in cancer chemotherapy, a bone marrow transplant is often required to restore a few of these cells to their body.

b. Difficult to recover rare adult stem cells and to keep them from differentiating in tissue culture. Dr. Irving Weissman's lab at Stanford was the first to recover adult stem cells (from bone marrow).

c. Incomplete knowledge of the trophic factors and tissue environment required for transplanted stem cells to perform correctly.

d. Surprises from some experiments, i.e. that stem cells for muscle implanted in muscle produce some cells that end up in the brain.

e. Ethical debate over whether a human embryo is a human life. Clearly the cells are alive ... but should excess embryos from in vitro fertilization be destroyed? Be frozen? Be donated to science?

Discussion

How do we balance the desire of those suffering from debilitating illness with the dignity of human life, even of an embryo? Will embryonic stem cell research lead inevitably to human cloning?

Target Diseases for Stem Cell Therapies

Spinal cord injury-grow new nerves

Parkinson's--implant embryo cells that can produce L-dopa in the brain

Juvenile diabetes-new islet cells

Leukemia-new immune system stem cells

Heart disease-cardiac stem cells to make new, healthy muscle

Cystic fibrosis-selected tissue repair, for example in the lining of the lungs, with healthy cells to relieve symptoms

Alzheimer's-new brain cells.

BIOPESTICIDES AND BIOTECHNOLOGY

For centuries, humans have improved crop plants through selective breeding and hybridization-the controlled pollination of plants. In more recent times, plant breeders created new varieties using chemicals or irradiation to provide unique traits in plants. Plant biotechnology is an form of plant breeding with one very important difference-plant biotechnology allows for the transfer of genetic information from species unrelated to the plant.

Traditional plant breeding involves the crossing of hundreds or thousands of genes, whereas plant biotechnology allows for the transfer of only one or a few desirable genes. This more precise science allows plant breeders to develop crops with specific traits and without relying on imprecise chemical mutation or random cross pollination that may include undesirable traits that have to be bred out of the new plant before it can be a commercially viable new variety.

Some of the beneficial traits in new plant varieties fight plant pests-insects, weeds and diseases-that can be devastating to crops. The pest fighting traits are pesticides as defined by the Federal Insecticide, Fungicide, and Rodenticide Act (FIFRA). The EPA regulates the use of pesticides under the authority of three federal statutes-FIFRA, the Federal Food, Drug, and Cosmetic Act (FFDCA), and the Food Quality Protection Act (FQPA). Federal oversight of pesticides is required to ensure that their use will not cause unreasonable adverse effects to human health or the environment. Other laws that may be taken into consideration when conducting risk assessments include the Endangered Species Act, the Migratory Bird Treaty Act and other environmental laws.

EARLY REGULATORY DEVELOPMENT FOR BIOTECHNOLOGY PRODUCTS

Responding to the rapid increase in the production of biotechnology products, there was a realization for the need for some sort of guidelines to ensure that public health and the environment are adequately protected from the potential risks of this technology. As products began moving from the laboratory toward the market, regulatory agencies realized that there should be regulatory mechanisms to ensure that these new products did not adversely affect public health or the environment.

To clarify regulatory jurisdiction over biotechnology products, the Administration established an interagency working group under the White House Cabinet Council on Natural Resources and the Environment in 1984. The working group's principle goal was to ensure the regulatory process adequately considered health and environmental safety consequences of the products of biotechnology as they move from the laboratory to the marketplace. Safety was not their only concern; however, the Council also emphasized the importance of not stifling innovation or enervating the competitiveness of the U.S. biotech industry. Thus, the interagency working group sought to establish a sensible framework that effectively protected human health and the environment while providing breathing room for a burgeoning industry.

Scientists also wanted the freedom and flexibility to engage in research and did not want Congress to pass unduly restrictive laws. In 1986, the Federal Government, through the Office of Science and Technology Policy issued the Coordinated Framework for the Regulation of Biotechnology. The Coordinated Framework concluded that the products of biotechnology do not differ fundamentally from unmodified organisms or from conventional products; that the product, rather than the process, should be regulated; that the regulations should be based on the end use of the product; and that review should be conducted on a case-by-case basis.

The Coordinated Framework states that no new statutes (laws) are needed to regulate biotechnology products, but that new regulations would be required. All regulations would go through a process of public notice, a public comment period, and consideration of those comments before final regulations were established. EPA is also required under FIFRA to take such proposed regulations through our outside scientific peer review committee called the Scientific Advisory Panel.

As part of a coordinated federal framework for biotechnology, biotech crops undergo a food safety and environmental approval process by the EPA, Food and Drug Administration (FDA) and the Department of Agriculture (USDA). The FDA's role (FFDCA), is to insure the safety of the food supply which is typically done by verifying that any new biotech foods are as safe as the foods already being consumed. The USDA (Federal Plant Pest Act-FPPA, Plant Quarantine Act-PQA and the Virus, Serum, Toxin Act-VSTA) is responsible for protecting US agriculture by assessing biotech plants for their potential to become a plant pest in the environment, while EPA's mandate is to protect human health and the environment by ensuring that a plant derived from biotechnology produces no unreasonable adverse effects.

EPA'S REGULATION OF BIOTECHNOLOGY PESTICIDE PRODUCTS

EPA regulates three major classes of biopesticides: biochemical pesticides, microbial pesticides, and plant-incorporated protectants (PIPs). Within the office Pesticide Programs (OPP), the Biopesticides and Pollution Prevention Division (BPPD) regulates natural and engineered microbial pesticides, PIPs and biochemical pesticides. Microorganisms can be genetically engineered to produce biochemicals used as pesticides. Microbial pesticides can be naturally occurring or genetically engineered. Plant-Incorporated Protectants are pesticidal substances that are produced in a living plant along with the genetic material necessary to produce the substance, where the substance is intended for use in the living plant. EPA includes the genetic material necessary to produce the substance in the definition of a plant-incorporated protectant because the genetic material introduced into the plant will ultimately result in a pesticidal effect. In addition, the inclusion of the genetic material may be responsible for the spread of the pesticidal trait in the environment to related plant relatives.

Under the Coordinated Framework, EPA proposed regulations for genetically engineered microbial pesticides and for plant-incorporated protectants. Regulations for notification prior to small scale field testing of engineered microbial pesticides were final in

1994. At the end of 1994, EPA proposed a set of regulations aimed at establishing the Agency's scope for PIPs and proposing to exempt several classes of PIPs which had very low risk. Most of these proposed regulations were final in 2002, although some of the exemptions are being further evaluated before the Agency makes a final decision.

Genetically Engineered Microbial Pesticides

Genetically engineered microorganisms are regulated using essentially the same data requirements used for naturally occurring microbial pesticides. (See 40 CFR part 158.740.) Some additional data may be required concerning the genetic engineering process used and the results from that process.

EPA requires notification prior to small scale field testing of genetically engineered microorganisms to allow EPA to determine if an Experimental Use Permit is needed. (See 40 CFR part 172 subpart C.) When testing10 acres or more, EPA requires an Experimental Use Permit before field testing naturally occurring or genetically engineered microorganisms. Under FIFRA, microbial biotech products, as with all other pesticides, must be evaluated for their risks and benefits. Before any registration is granted, OPP considers such issues as potential adverse effects to non-target organisms, environmental fate of the microorganism, and the potential pathogenicity and infectivity of the microorganism to humans.

Plant-Incorporated Protectants

All plants produce some substances which act to repel certain insects or kill various pathogens such as bacteria and fungi; plants can also have other mechanisms such as hairy leaves or other structure to help ward off pests. Plants which have more of these natural protections are less susceptible to pests than plants which have fewer of them.

In the 1994 policy statement and proposed rules for what are now called PIPs (plant-incorporated protectants), EPA explained that substances plants produce for protection against pests are pesticides under FIFRA, if humans intend to use these substances for pesticidal purposes. In 1994, EPA called these substances, produced and used in living plants, along with the genetic material necessary to produce them, "plant-pesticides." With issuance of the policy statement and proposed rules, EPA began registering plant-pesticides under FIFRA and regulating residues of plant-pesticides under FFDCA. The name plant-incorporated protectants or PIPs was selected after EPA took comments on a substitute name for plant-pesticides as requested in the public comments in response to the 1994 proposed rules.

In 2001, EPA published final rules exempting from FIFRA requirements (except for an adverse effects reporting requirement pesticidal substances produced through conventional breeding of sexually compatible plants. EPA also exempted residues of these pesticidal substances and of all nucleic acids that are part of a PIP from FFDCA pesticide residue requirements. These residues of pesticidal substances were exempted based on a long history of human dietary exposure to these naturally occurring plant compounds, and epidemiological studies showing the health benefits of consuming foods which can contain low levels of these substances. Residues of nucleic acids that are part of PIPs were exempted as nucleic acids are common to all life forms, have always been present in human and domestic animal food, and are not known to cause adverse health effects when consumed as part of food.

PIPs moved into plants from other organisms, including from plants not sexually compatible with the recipient plant, continue to be regulated. EPA believes this oversight is appropriate because of the new exposure to the PIP now expressed in the new host plant. An example of this type of PIP is the insecticidal protein from the bacterium Bacillus thuringiensis introduced into a plant for protection against lepidopteran (caterpillar) pests. EPA also proposed in 1994 to exempt three other categories of PIPs. However, in 2001, EPA decided to solicit further guidance from the public on the three proposed exemptions because of the wide range of varied comments received on the proposals. In addition, EPA placed in the public record for the rules and solicited comment on the information, analyses and conclusions of the 2000 National Academy of Sciences study entitled "Genetically Modified Pest-Protected Plants: Science and Regulation" that recommended EPA reconsider its proposed categorical exemption of viral coat proteins, PIPs which act by primarily affecting the plant, and PIPs produced by using biotechnology to move genes between sexually compatible plants.

In general, the data requirements for PIPs are based on those for microbial pesticides as, up to now, the PIP products have come from microorganisms. The exact data requirements for each product have been developed on a case by case basis. All of the products EPA has seen to date have been proteins, either related to plant viruses or based on proteins from the common soil bacteria Bacillus thuringiensis also known at Bt. The general data requirements include product characterization, mammalian toxicity, allergenicity potential, effects on non-target organisms, environmental fate, and for the Bt products, insect resistance management to product from losing use of both the microbial sprays and the Bt PIPs. In December 1999 and June 2000 EPA asked the Scientific Advisory Panel (SAP) to reevaluate the data requirements for protein PIPs. The Agency presentations and the SAP reports are available on the website listed below. EPA has made adjustments to our data requirements based upon the SAP's recommendations.

EPA's recent reevaluation of the Bt PIPs shows that generally, these products reduce the use of more toxic insecticides. This is most obvious for Bt cotton. Traditionally, cotton is one of the most frequently treated crops, often with organophosphate compounds. The use of Bt cotton greatly reduces the use of these toxic compounds. New products currently being tested under Experimental Use Permits to control corn rootworm have the potential to reduce reliance on the millions of pounds of organophosphates and carbamates currently used to control this insect.

One should not, however, confuse support with advocacy. As a regulatory agency, the EPA maintains a neutral position regarding biotechnology. We require appropriate studies to determine if there is a risk and the studies must conform to established standards including being conducted by labs that are subject to inspection. The studies enable the agency to conduct risk assessments that are made available to the public. Additionally, the Agency strives to have a transparent review process that enables all stakeholders views to be heard. EPA maintains a public docket of all studies and the Agency's review of the studies; we subject our analyses to the Scientific Advisory Panel (SAP); and the Agency holds public hearings, workshops, and seminars to provide the public with information and to obtain input on our regulatory decisions.

All these efforts were designed to assess and regulate the human health and environmental impacts of biotechnology products, and to assure that the decisions taken by the Agency were based on the most current health and ecological data that provide consistency, effectiveness, and flexibility. As the science of biotechnology progresses, we anticipate that the types of products will also change. These changes will determine how we review these ingredients and the types of data that will be required. In the future, it is possible that compounds similar to protective compounds already produced by plants, such as phenols, ketones or aldehydes, will be used as PIPs. These types of compounds will require a very different set of data than that required to determine the safety of proteins. EPA will rely on the SAP and similar scientific bodies to guide the Agency on the appropriate data requirements. EPA remains fully committed to assuring that the review, assessment and registration of biotechnology products meet the stringent standards required by FIFRA and the FFDCA, and are fully protective of public health and the environment. It is EPA's goal to make our regulatory process and decisions within a sound and transparent process framework tuned into the most recent and scientifically sound information.

Herbicide Tolerant Plants

The EPA regulates the herbicides used on herbicide tolerant plants, but the regulation of the plants and the food crops is done by USDA and FDA, respectively. EPA does establish tolerances (legal limits of pesticide residue) for the herbicide used to control weeds in the crop fields. In addition, EPA and USDA have a Memorandum of Understanding to work together on resistance management for herbicide tolerant crops.

DATA REQUIREMENTS FOR THE REGISTRATION OF PLANT-INCORPORATED PROTECTANTS

Product Characterization

Identification of the new Plant-Incorporated Protectants (PIP) character added to the plant generally follows guidance developed by the EPA, Canadian Food Inspection Agency (CFIA) and the USDA's Animal and Plant Health Inspection Service (APHIS). This includes a description of the source of the pesticidal gene (promoters, enhancers, open reading frame and closing and release signals for the pesticidal trait) and identifies the source of any other traits introduced such as selection markers (herbicide resistance, antibiotic resistance or a phenotypic trait such as cells turning blue when introduced $-glucuronidase is present). This information is given as a plasmid map of the vector DNA and includes the exact DNA sequence for the introduced traits. The discussion also includes the biology of the source organism(s) and describes any hazards associated with source organism(s) such as pathogenicity or toxin production. Rationale is also given as to why this trait was selected and any changes to the actual DNA sequence of the introduced gene(s) are described.

A description of the recipient plant is provided including the general biology and use as a crop. The details expected to be covered would include the possible production of toxins or anti-nutrients by the plant, major insect pests, weeds and diseases of the crop, reproductive biology, and presence of wild or weedy relatives of the crop in the United States.

Trait introduction is discussed to provide information about the plant cell culture and selection/regeneration technique used to produce the genetically engineered plant. Confirmatory data show what part of the vector DNA is actually incorporated into the plant genome. This consists of southern blot analysis for the trait's gene and also gives information about the possibility of multiple copies of the gene(s) being incorporated. A southern blot analysis is a DNA detection assay based on the high binding affinity of the two complimentary strands of DNA for each other. Using a radioactive or other labeled form of one of the complimentary strands and restriction endonuclease enzymes which cut the DNA at specific locations, information about the status and insertion of the introduced traits in the transformed plant can be surmised.

The stability and inheritance of trait is examined to determine any linkage of the introduced trait(s) and if there is more than one site of incorporation for the trait(s). The presence and performance of the trait over several plant generations is examined for determining stability of trait expression.

Protein characterization and expression data provide biochemical information about the actual expressed protein in the plant and its concentration in various tissues. This includes the amino acid sequence, the activity of the protein (usually information about the range of susceptible species) and identification of the expressed protein. Protein identification usually includes SDS-PAGE analysis, immunological recognition in an ELISA or western blot assay and N-terminal amino acid sequencing.

For the proteins reviewed to date by the EPA, there has also been information to indicate if the protein was modified when produced in the plant versus its form in the original donor organism. For traits derived from bacteria, there is a possibility that the protein was modified after translation in the plant, for example by adding extra sugar residues to the protein (glycosylation), which does not occur in the source bacterium.

Protein glycosylation is a concern if it changes the stability or activity of the expressed protein and data from the microbial source is being used to determine safety. Expression data are provided to determine maximum exposure levels for the PIP in several plant tissue (stem, leaf, root, flower, pollen, etc). Expression levels are cogent for human health and environmental hazard assessment as well as insect resistance management.

Mammalian Toxicity

The mammalian toxicity data examined is guided by the fact that most PIPs seen to date are proteins. Given this fact a special set of data is generally considered to affirm the assumption that the introduced protein behaves like other dietary proteins. In vitro digestibility data is often considered part of the biochemical characterization of the expressed protein but may also address toxicity issues. The rationale behind its use is to determine if the expressed protein degrades in the presence of acid, heat or simulated gastric and intestinal digestive fluids as would be expected of a normal dietary protein. The stability to digestive fluids, acid or heat also is one of the complex of characteristics that is examined to determine if a protein with no dietary history has potential to be a food allergen. If a protein is not rapidly broken down by digestive fluids, it may have a longer time period to interact with the gastrointestinal mucosa and possibly induce an allergic reaction in susceptible individuals.

Acute oral toxicity is examined since the primary route of exposure is dietary for PIPs. The study is done with a gavage (forced oral) administration of a maximum hazard dose (2-5gm/ kg body weight) and the animals are followed for 14 days of observation for clinical signs. At the end of 14 days the animals are sacrificed and the internal organs are examined by gross necropsy. The test substance used is the protein expressed from the gene introduced into the plant or the same protein produced from an alternate source. If the protein has no unusual persistence in digestive fluids and shows no toxicity in the acute oral toxicity tests, there is no reason to believe that the introduced protein would behave any differently than any other dietary protein and longer term toxicity testing would not be justified.

Test substance equivalence may be required if an alternate source for the producing test material has been employed. In all the PIPs examined to date, there has never been sufficient expression of the protein PIP in the plant itself to provide adequate purified material to supply an acute oral toxicity test. Therefore, to provide adequate protein test material for the study, an alternate production system for the protein is employed. This has been either the original source bacterium Bacillus thuringiensis or an industrial fermentation strain like a laboratory strain of Escherichia coli.

In order to be able to utilize this alternate source, however, a series of tests similar to those done for product characterization need to be performed using both the plant expressed protein and the bacterial protein to insure these two are equivalent. These tests include the bioactivity of the protein (e.g., LD50 determinations against a range of insects), SDS-PAGE analysis, immunological recognition in an ELISA or western blot assay and possible post-translational modifications for the plant expressed material.

Amino acid sequence homology is examined for the introduced protein to determine if there are any similarities between the introduced protein and known protein toxins or allergens. This analysis is done by both a whole sequence homology comparison and also a step wise eight contiguous amino acid sequence comparison. The eight amino acid fragment analysis is done for comparison to known allergens since this is believed to be the smallest stretch of amino acids that can be recognized by an antibody. Since antibodies are intimately involved in the aberrant response causing food allergies, this analysis is critical for examining proteins with no previous dietary exposure for their potential to be related to a food allergen.

GENE FLOW ASSESSMENT FOR PLANT-INCORPORATED PROTECTANTS

The movement of transgenes from the host plant into weeds and other crops has been a significant concern for EPA due to the possibility of novel exposures to the pesticidal substance. This concern has been considered for each of the Bt plant-incorporated protectants (PIPs) currently registered.

The Federal Insecticide, Fungicide and Rodenticide Act directs the U.S. EPA to examine all potentially adverse environmental impacts, including those which may arise from gene flow of PIPs to wild or feral populations of sexually compatible plants. In addition to this mandate, the Federal Food, Drug, and Cosmetic Act requires the issuance of a food tolerance or exemption from the requirement of a tolerance for all pesticidal substances that may enter the food supply whether through seed mixing or cross pollination.

To date, PIPs in three crop species have been registered by the Agency and all have received exemptions from the requirement of a tolerance. Bt Corn, cotton and potato were reviewed for their potential to hybridize with wild and feral relatives of sexually compatible plants in the U.S., its territories and possessions.

EPA has determined that with the conditions of registration in place there is no significant risk of gene capture and expression of any Bt endotoxin by wild or weedy relatives of corn, cotton, or potato in the U.S., its possessions or territories. In addition, the USDA/APHIS has made this same determination under its statutory authority under the Plant Pest Act.

The Bt corn and potato PIPs that have been registered to date have been expressed in agronomic plant species that, for the most part, do not have a reasonable possibility of passing their traits to wild native plants.

Most of the wild species in the United States cannot be pollinated by these crops (corn and potato) due to differences in chromosome number, phenology and habitat. There is a possibility, however, of gene transfer from Bt cotton to wild or feral cotton relatives in Hawaii, Florida, Puerto Rico and the U.S. Virgin Islands. Where feral populations of cotton species similar to cultivated cotton exist, EPA has prohibited the sale or distribution of Bt cotton in these areas. These containment measures prevent the movement of the registered Bt endotoxin from Bt cotton to wild or feral cotton relatives.

EPA's evaluation of the possibility of movement of Bt transgenes into weedy relatives of potato, corn, and cotton are presented below:

Potato (Solanum tuberosum)

EPA has reviewed the potential for gene capture and expression of Bt plant-incorporated protectants (only Cry3A has been introduced into potato) by wild or weedy relatives of cultivated potato in the United States, its possessions or territories. Based on data submitted by the registrant and a review of the scientific literature, EPA concluded that there is no foreseeable risk of unplanned pesticide production through gene capture and expression of the Colorado potato beetle control protein (Cry3A) in wild potato relatives in the U.S.

Tuber-bearing Solanum species, including S. tuberosum. S. tuberosum cannot hybridize naturally with the non-tuber bearing Solanum species in the U.S. Three species of tuber-bearing wild species of Solanum occur in the United States: Solanum fendleri, Solanum jamesii, and Solanum pinnatisectum. But, successful gene introgression into these tuber-bearing Solanum species is virtually excluded due to constraints of geographical isolation and other biological barriers to natural hybridization.

These barriers include incompatible (unequal) endosperm balance numbers that lead to endosperm failure and embryo abortion, multiple ploidy levels, and incompatibility mechanisms that do not express reciprocal genes to allow fertilization to proceed. No natural hybrids have been observed between these species and cultivated potatoes in the U.S.

Corn/Maize. (Zea mays)

EPA has reviewed the potential for gene capture and expression of the BtCry endotoxin protein in corn plants, by wild or weedy relatives of maize in the United States. EPA believes there is no significant risk of gene capture and expression of any of the Cry endotoxins by wild or weedy relatives of maize in the United States because extant populations of sexually compatible species related to Zea mays are not present in the continental United States or its territories and possessions. Zea mays is a wind-pollinated species, and the presence of spatially separate tassels (male flowers) and silks (female flowers) encourages outcrossing among nearby plants. Maize cultivars and landraces are known to be interfertile to a large degree.

Recent studies have indicated that cross-pollination at 100 ft. from the source of genetically modified maize was 1 % and this proportion declined exponentially to 0.1 % at 130 ft and further declined to 0.03 % at the farthest distance measured (160 ft). For production of Foundation Seed, a distance of 660 ft has been required to ensure separation of pollen types. Additionally, the relatively large size of corn pollen as compared to other grass species and the short time span that corn pollen remains viable (i.e., typically less than 60 minutes) under natural conditions both preclude long distance transfer for purposes of outcrossing. Under conditions of high temperature and desiccation, corn pollen longevity is measured in minutes. These conditions may even destroy the anthers before any viable pollen is shed. More moderate conditions can extend the field life to hours.

For transformed plants to become weedy escapes as a result of the genetic modification (i.e., expression of Bt endotoxins that protect plants from insect damage), they would need to inherit and express many other unrelated traits that provide selective advantage to a weedy growth habit (e.g., large numbers of easily dispersed seeds, propensity to grow on disturbed ground, vegetative propagation, seed dormancy, etc.). These traits do not exist within the corn complement of genetic characters, a species that has been selected for domestication and cultivation under conditions not normally found in natural settings. The presence of a large cob or ear that does not shatter as the bearer of seeds severely limits the dispersing abilities of corn and it has been theorized that in the species as we know it would die out in a few generations due to competition amongst seedlings germinating from the cob.

Transformation of corn to express Bt endotoxin does not alter the ability of corn to outcross with teosintes (Zea mays ssp. mexicana, Z. mays ssp. parviglumis, Z. luxurians, Z. perennis, Z. diploperennis) or Tripsacum species. Teosintes exist as special plantings (e.g., in research plots, botanical gardens, and greenhouses) and some are used to a small extent as forage crops in the western United States. Many native teosintes in Mexico, El Salvador, Guatemala, Nicaragua and Honduras are interfertile with corn to varying degrees and have been known to produce viable seedlings.

Despite having coexisted and co-evolved in close proximity to corn in the Americas over thousands of years, however, corn and teosintes maintain distinct genetic constitutions even with this sporadic introgression.

Given the cultural and biological relationships of various teosinte species and cultivated corn over the previous millennia, it appears that gene exchange has occurred (based largely upon morphological characters) between these two groups of plants and that no weedy types have successfully evolved as a result. More recent cytogenetic, biochemical and molecular analysis has indicated that the degree of gene exchange is far less than previously thought and evidence for gene introgression into teosinte from corn may be considered as circumstantial at present.

The teosintes retain a reduced cob-like fruit/inflorescence that shatters more than cultivated corn, but still restricts the movement of seeds as compared to more widely dispersed weedy species. Hence, the dispersal of large numbers of seeds, as is typical of weeds, is not characteristic of teosintes or corn. In their native habitat, some teosintes have been observed to be spread by animals feeding on the plants. Teosintes and teosinte-corn hybrids do not survive even mild winters and would not propagate in the U.S. Corn Belt. Additionally, some types have strict day length requirements that preclude flowering within a normal season (i.e., they would be induced to flower in November or December) and, hence, seed production under our temperate climate.

Based on the ability of corn to hybridize with teosintes, the results of previous genetic exchange amongst these species over millennia, and their general growth habits, any introgression of genes into wild teosinte from Zea mays is not considered to be a significant agricultural or environmental risk.

The growth habits of teosintes are such that the potential for serious weedy propagation and development is not biologically plausible in the United States.

Sixteen species of Tripsacum are known worldwide. Most of the 16 different Tripsacum species recognized are native to Mexico, Central and South America, but three occur within the U.S. The Manual of Grasses of the United States reports the presence of three species of Tripsacum in the continental United States: T. dactyloides, T. floridanum and T. lanceolatum. Of these, T. dactyloides, Eastern Gama Grass, is the only species of widespread occurrence and of any agricultural importance. It is commonly grown as a forage grass and has been the subject of some agronomic improvement (i.e., selection and classical breeding). T. floridanum is known from southern Florida and T. lanceolatum is present in the Mule Mountains of Arizona and possibly southern New Mexico.

For the species occurring in the United States, T. floridanum has a diploid chromosome number of 2n = 36 and is native to Southern Florida. T. dactyloides includes 2n = 36 forms which are established in the central and western U.S., and 2n = 72 forms which extend along the Eastern seaboard and along the Gulf Coast from Florida to Texas, but which have also been found in IL and KS; these latter forms may represent tetraploids (x = 9 or 18). T. lanceolatum (2n = 72) occurs in the Southwestern U.S. Eastern Gama Grass (T. dactyloides) differs from corn in many respects, including chromosome number (T. dactyloides commonly n = 18; Zea mays n = 10). Many species of Tripsacum can cross with Zea, or at least some accessions of each species can cross, but only with difficulty and the resulting hybrids are primarily male and female sterile. T. dactyloides, is considered by some to be an ancestor of Zea mays or cultivated corn, while others dispute this, based largely on the disparity in chromosome number between the two species, as well as radically different phenotypic appearance.

Albeit with some difficulty, hybrids between the two species have been made. In most cases these progeny have been sterile or viable only by culturing with in vitro 'rescue' techniques.

Relatively few accessions of T. dactyloides will cross with corn and the majority of progeny aren't fertile or viable even in those that do. In controlled crosses, if the female parent is corn, there is a greater likelihood of obtaining viable seed. When these hybrids have been backcrossed to corn in attempts to introgress Tripsacum genes for quality enhancement or disease resistance, the Tripsacum chromosomes are typically lost in successive generations.

Even though some Tripsacum species occur in areas where corn is cultivated, gene introgression from corn under natural conditions is highly unlikely, if not impossible. Hybrids of Tripsacum species with Zea mays are difficult to obtain outside of the controlled conditions of laboratory and greenhouse. Seed obtained from such crosses are often sterile or progeny have greatly reduced fertility. Approximately 20% of corn-Tripsacum hybrids will set seed when backcrossed to corn, and none are able to withstand even the mildest winters. The only known case of a naturally occurring Zea-Tripsacum hybrid is a species native to Guatemala known as Tripsacum andersonii. It is 100 % male and nearly 99% female sterile and is thought to have arisen from an outcrossing to a teosinte, but the lineage is uncertain. Zea mays is not known to harbor properties that indicate it has weedy potential and other than occasional volunteer plants in the previous season's corn field, corn is not considered as a weed in the U.S. The risk of Tripsacum/corn hybrids forming in the field is considered minimal. Tripsacum species are perennials and seem more closely related to the genus Manisurus than either to corn or teosinte.

Since both teosinte and Tripsacum are included in botanical gardens in the U.S., the possibility exists (although unlikely) that exchange of genes could occur between corn and its wild relatives. EPA is not aware, however, of any such case being reported in the United States. Gene exchange between cultivated corn and transformed corn would be similar to what naturally occurs at the present time within cultivated corn hybrids and landraces. Plant architecture and reproductive capacity of the intercrossed plants will be similar to normal corn, and the chance that a weedy type of corn will result from outcrossing with cultivated corn is extremely remote. Like corn, Zea mays ssp. mexicana (annual teosinte) and Zea diploperennis (diploid perennial teosinte) have 10 pairs of chromosomes, are wind pollinated, and tend to outcross, but are highly variable species that are often genetically compatible and interfertile with corn.

Zea perennis (perennial teosinte) has 20 pairs of chromosomes and forms less stable hybrids with corn. Corn and compatible species of teosinte are capable of hybridization when in proximity to each other. In Mexico and Guatemala, teosintes exist as weeds around the margins of corn fields. The F1 hybrids have been found to vary in their fertility and vigor. Those that are fertile are capable of backcrossing to corn. Except for special plantings, however, teosinte is not present in the U.S. or its territories. Its natural distribution is limited to Mexico, Honduras, Nicaragua and Guatemala. Tripsacum/corn hybrids have not been observed in the field, but have been accomplished in the laboratory using special techniques under highly controlled conditions.

Cotton. (Gossypium spp.)

EPA has reviewed the potential for gene capture and expression of the Cry1Ac endotoxin in cotton by wild or weedy relatives of cotton in the United States, its possessions or territories. There is a possibility for gene transfer in locations in Hawaii, Florida, Puerto Rico and the U.S. Virgin Islands where wild or feral cotton relatives exist. Therefore, EPA required stringent sales and distribution restrictions on Bt crops within these areas to preclude hybridization of the crop with sexually compatible relatives.

There are four species of cotton, Gossypium, in the United States. Two of them, Gossypium hirsutum (upland cotton) and Gossypium barbadense (sea island cotton, pulpulu haole, Pima), are used commercially and escaped plants can be found growing in the wild in climates where they can survive the winter. In addition, two native wild species of Gossypium occur in the United States: G. thurberi Todaro and G. tomentosum Nuttall ex Seeman. G. thurberi Todaro (Thurberia thespesiodes Gray) occurs in the mountains of Southern Arizona and northern Mexico at 2,500 to 5,000 feet (rarely at 7000 feet), and is rather common on rocky slopes and sides of canyons in late summer and autumn. Any gene exchange between plants of Gossypium hirsutum and Gossypium thurberi, if it did occur, would result in triploid (3x=39 chromosomes), sterile plants because G. hirsutum is an allotetraploid (4x = 52 chromosomes), and G. thurberi is a diploid (2x = 26 chromosomes). Such sterile hybrids have been produced under controlled conditions, but they would not persist in the wild; in addition, fertile allohexaploids (6x = 78 chromosomes) have not been reported in the wild.

The second wild native species, Gossypium tomentosum, occurs in Hawaii on the six islands of Kahoolawe, Lanai, Maui, Molokai, Nihau and Oahu. Upland, Hawaiian and sea island cotton are all tetraploids (4x = 52) that can crossbreed. Introgression has been claimed for what one author considered hybrid populations of G. barbadense X G. tomentosum. As G. tomentosum may bloom at the same time as domestic cotton, there is no guarantee of either geographic or temporal isolation. For these reasons, EPA imposed stringent sales and distribution restrictions on the registration for cotton expressing the Cry1Ac delta endotoxin grown in Hawaii. The Agency required the following labeling statement to mitigate the potential for the Cry1Ac gene to move from cultivated cotton to G. tomentosum:

"Not for commercial sale or use in Hawaii. Test plots or breeding nurseries established in Hawaii must be surrounded by either 12 border rows of non-cotton if the plot size is less than 10 acres or 24 border rows if the plot is over 10 acres and must not be planted within 1/4 mile of Gossypium tomentosum."

The inability of plants or seeds of either of G. hirsutum or G. barbadense to survive freezing temperatures restricts their persistence as perennials or recurrent annuals to tropical areas. Feral G. hirsutum occurs in parts of southern Florida in the Everglades National Park and the Florida Keys. Cotton is not grown commercially in these areas at this time (cultivated cottons are found in the northernmost portions of the state), but the containment provisions of the initial registration must continue for areas in Florida where feral cotton occurs. Wild cotton is a potential concern as it may increase the spread of resistance in Florida (with intensive vegetable production). EPA imposed sale and distribution restrictions on Bt cotton in Florida, restricting its use to those sites North of Tampa (Route 60).

The Agency is satisfied that the planting restrictions on Bt cotton (i.e., no Bt cotton south of Tampa) will mitigate concerns for gene transfer to wild cotton:

"Do not plant in Florida South of Tampa, Route 60."

A restriction on the planting of Bt cotton in Florida south of Route 60, near Tampa, precludes any chance of outcrossing with feral Gossypium spp. due to the extremely large distance between any commercial plantings and wild populations in the extreme south of the state. Previously there were no restrictions on commercial planting of Bt cotton in the Virgin Islands or Puerto Rico, both of which are known to have extant populations of wild G. hirsutum or G. barbadense or hybrids of the two species. In the former case, no commercial plantings of cotton or test plots of Bt cotton are planned for the U.S. Virgin Islands, hence, gene flow is not a concern there.

Planting of Bt cotton should be restricted, however, if prevention of gene flow is desired. Puerto Rico has seen use as a winter nursery for some commercial cotton breeding and the populations of wild Gossypium are apparently commonly distributed throughout the island.

Similar restrictions, as imposed for Hawaii, are appropriate for this situation based upon the distribution of wild populations around the island and the propensity of insect pollinators to move substantial differences.

In both cases, monitoring of native populations of established Gossypium spp. may be necessary to assess the efficacy of this isolation procedure for Bt cotton. This would entail monitoring of wild populations for evidence of gene introgression through PCR or similarly sensitive methods. Alternatively, the absolute restriction of planting Bt cotton in Puerto Rico and the U.S. Virgin Islands, would of course, alleviate any concerns over gene flow.

Given that the distribution of feral Gossypium spp. in Puerto Rico is not well documented, the mapping of native populations with related distance to breeding nurseries of Bt cotton may be sufficient to indicate the probability of cross pollination. If any populations of feral Gossypium spp. are eliminated over time, this could potentially change the proximity to research areas and influence the isolation requirements with respect to gene flow issues.

EPA analyzed the possibility of gene flow from transgenic cultivars expressing Bt to wild native plants which acquire the Bt genes through cross pollination as part of its risk assessment of Bt potato, Bt corn, and Bt cotton as summarized above.

EPA has concluded that there is no significant risk of gene capture and expression of the Bt toxin(s) by wild or weedy relatives of corn, cotton, or potato in the U.S., its possessions or territories. The creation of "Bt-enhanced weeds" from Bt gene movement from Bt corn or Bt potato is only a remote possibility in the U.S. For Bt cotton, where the possibility of gene movement may exist in certain geographically distinct areas, EPA has mitigated the potential for such movement by imposing strict geographic restrictions on the sale and distribution of Bt cotton.

ECOLOGICAL NON-TARGET ORGANISM RISK ASSESSMENT

Non-Target Organism Hazard Assessment Process

Hazard to non-target organisms is a function of toxicity AND exposure. Non-target organisms that are not exposed to a toxic substance are not at risk. Organisms that may be exposed to limited amounts of a toxic substance may also not be at risk.

Therefore, the risk assessment process depends on determination of the toxicity of a given substance and the degree of environmental exposure of non-target species. The primary issue which defines the type of toxicity data needed for a risk assessment of PIPs is that the pesticidal substance is contained within the plant parts thus resulting in minimal exposure to non-target organisms. This exposure scenario is quite different from spray applications of pesticides. Exposure of non-targets to plant-incorporated protectants would occur primarily when wildlife feed on the pesticidal plants, on exposure to dispersed pollen or if sexual transfer of the new trait(s) to non-target wild/weedy relatives by cross-pollination takes place. Therefore, each risk assessment is made from an analysis of the nature of the gene being introduced, the plant receiving the gene, the environment where the plant will be grown and the species susceptible to the effects of the introduced gene. This amounts to a case-by-case analysis. The Basic Issues In Non-target Organism Risk Analysis are as follows:

1. The non-target organism endpoints of concern for PIPs are initially based on containment of the gene product to the modified plant (except for pollen dissemination).
2. This containment limits the exposure of the gene product to non-target organisms which are expected to interact with the plant.
3. Where exposure occurs, the possible toxicity to non-target organisms is evaluated through non-target wildlife testing.
4. If toxicity to the non-target wildlife in question is observed, then the amount of exposure is determined to ascertain if adverse effects to any non-target could occur under field conditions.

In addition, the concentration of the active ingredients in plant tissues, soil residues and their degradation rates, are also measured to more accurately determine exposure of non-target organisms. The Agency believes that the single species laboratory toxicity testing at doses 10X to 100X the levels encountered in the environment (using a "safety factor") represents a reasonable approach to evaluating risk related to the use of biopesticides, and is one in which negative results at high exposures would allow a high degree of confidence in the safety of the test agents. Where testing shows a hazard at doses approaching the expected environmental concentrations, a determination of the LD50, LOEC and NOEC is required. If necessary, higher tier testing (to assess population level effects) is performed.

Non-Target Organism Data Requirements

The following basic ecological effects testing requirements on representative non-target terrestrial and aquatic species listed in the OPPTS Harmonized Pesticide Test Guidelines are addressed by submission of data or waiver requests with credible justification:

The testing required by the OPPTS Harmonized Pesticide Test Guidelines is as follows:

Tier I:

- Avian oral toxicity test (on upland game bird and waterfowl species) (A 6 week broiler dietary study for chronic exposure effects has also been added)
- Wild mammal oral toxicity test (rodent species)
- Freshwater fish oral toxicity test (on cold water and warm water species) (Including farmed fish feeding studies for chronic exposure effects)
- Freshwater invertebrate testing (on Daphnia or aquatic insect species)
- Estuarine and marine animal testing (grass shrimp and fathead minnow species)
- Non-target plant studies (for terrestrial, aquatic and out-crossing issues)
- Honey bee testing (for larval and adult bee toxicity)
- Non-target insect testing (on predators and parasites, most commonly green lacewing larvae, the ladybird beetle, a parasitic wasp and soil invertebrates).
- Tritrophic testing of selected beneficial insects is also being submitted by the registrants.
- Terrestrial environmental expression testing (which is a determination of degradation rates of the proteins in soil) (and)
- Plant tissue expression data and degradation rates

If the results of Tier I testing show adverse non-target species effects at field use rates, then testing of additional species and/ or testing at a higher Tier level is required:

Tier II: Freshwater and marine or estuarine environmental expression testing (aquatic environmental fate).

Tier III: Chronic, reproduction, life cycle and population effects (and host range) testing.

[If the results from environmental fate studies show a plant protein that is toxic to non-target species persists in the environment at significant levels, Tier III studies are designed to show effects of chronic exposure to these levels on selected fish and wildlife species. Tier III studies are also used to determine non-target effects of plant proteins designed to inhibit insect molting, reproduction, disease resistance and similar properties]

Tier IV: Simulated or actual field testing (however, field scouting for non-target insect abundance is currently recommended as a Tier I test)

Tier IV studies are used to determine if there is a noticeable shift in wildlife populations under field use conditions.

OPP recognizes the potential value of Tier IV field tests as an additional check on the presence of ecosystem effects. Therefore, field data are being gathered concurrently with full-scale efficacy testing during the product development stage.

This provides the opportunity to evaluate pesticidal effects (both direct and indirect) on a much broader spectrum of non-target species under more natural exposure conditions than is possible in single species no-choice laboratory feeding studies.

Non-target insect field studies also address the long term use effects at the ecosystem level.

Candidate Test Organisms

An additional factor in determining the extent of testing necessary for risk assessment is the degree of pest species specificity shown by the protein in question.

This is of primary importance in assessing ecological risk. Most protein plant-incorporated protectants produce adverse effects against a specific class of target species. Careful scientific consideration on a case-by-case basis is given to the selection of non-target species to be tested in order to include species that are most likely to be susceptible.

Selection of the test species for any specific plant construct is done jointly by the Agency and the registrant during pre-registration conferences. Rationale for selection is discussed and the final test species list may be smaller or greater than the guideline data requirements.

The selection of test species is limited by their availability, the ability to rear them in captivity-in sufficient quantities for testing, or their ability to survive captivity without unacceptable mortality levels.

Endangered and threatened species hazard is addressed by extrapolating toxicity from related species, and if a hazard is identified, after consultation with the Fish and Wildlife Service possible risk is mitigated by preventing field exposure.

Test Substances and Dosing

Wherever possible, the whole modified plant tissue or pollen is used as the dosing substance. The use of pure test substance alone may be inadequate since it does not test for inadvertent, possibly harmful changes in the plant tissue itself (i.e. unintended effects).

In addition to the general data requirements discussed above, data derived from additional tests may be required by the Agency in order to make judgments regarding safety to non-target organisms on a case-by-case basis. Such data may also be required where special problems with Tier I testing are encountered.

Test methods will usually be derived from protocols already described or cited in OPPTS Harmonized Guidelines, or other sources, such as the OECD Guidelines, or new protocols may be developed on a case-by-case basis.

ENVIRONMENTAL FATE

Fate and Environmental Effects of Transgenic Bacillus Thuringiensis d-Endotoxins in the Soil

Soil organisms may be exposed to d-endotoxins from current transgenic crops by exposure to roots, incorporation of above ground plant tissues into soil after harvest, or by pollen deposited on the soil.

Root exposure may occur by feeding on living or dead roots or, theoretically, by ingestion or absorption after secretion of d-endotoxin into the soil.

In addition, evidence suggests that some soil components, e.g. clays and humic acids, bind d-endotoxins in a manner that makes them recalcitrant to degradation by soil microorganisms, but without eliminating their insect toxicity. Therefore, exposure to d-endotoxin bound to soil particles may also be a route of exposure for some soil organisms.

Experiments addressing the amounts and persistence of d-endotoxins in the soil have been submitted and reviewed for the current registrations. A number of publications in the scientific literature have also addressed the degradation of Cry proteins in the soil. These experiments consist of the incorporation of purified d-endotoxin or transgenic plant material in soil in a laboratory setting. Cry protein half-life studies were submitted for registration for corn containing Cry1Ab, and published studies were available for Cry1Ac cotton. Cry1Ab produced estimated degradation rates of 1.6 days for the Cry protein as expressed in transgenic corn tissue and 8.3 days for the purified protein (Sims and Holden 1996).

Data produced by Monsanto for Cry1Ac protein and transgenic Cry1Ac in cotton give degradation rates of approximately 9-20 days for the purified protein, and 41 days for the protein in cotton tissue.

Published data for Cry1Ab or Cry1Ac in cotton tissue or as purified protein produced degradation rates of 2.2 to 46 days, where measurable (in 4 of 11 experiments), with degradation rates in transgenic tissue shorter than for purified protein in two of three experiments (Palm et al. 1994).

Degradation rates of purified Cry1Ac in two different non-sterile soils were 22 d and 40 d (Palm et al. 1994). None of the studies have been performed under field conditions, although most have used field soil in laboratory microcosms.

Several studies indicate that Cry proteins bind to clays and humic acids (Crecchio and Stotzky 1998, Koskella and Stotzky 1997, Tapp and Stotzky 1995, Tapp and Stotzky 1998, Stotzky 2000a). The results of these studies suggest that this binding slows the rate of microbial degradation of these toxins compared to when these soil components are not present (Stotzky 2000a).

However, this protection is not absolute, since degradation does in fact occur under several experimental conditions. Several factors influence either the affinity of binding or the rate of degradation. In particular, pH near neutrality generally substantially increases degradation.

At pH above 5.8 to near neutrality, degradation of Cry protein bound to clay minerals in soil was much faster than degradation at pH 4.9-5.0 (Tapp and Stotzky 1998). For example, it was found (Tapp and Stotzky 1998) that Cry toxin added to nonsterile soil containing kaolinite or montmorillonite showed little degradation even after around six months at lower pH (pH~5), while substantial degradation occurred over this time period at higher pH.

Corn does not grow well below ~pH 5.6 (Aldrich et al. 1975), and therefore most corn growing soils are expected to be at a higher pH. Potato prefers acid soils (Smith 1977), and the optimum range is pH 5.0-6.5 (Ware and McCollum 1980). The optimal range for cotton is pH 6.0-6.5 (Donahue et al. 1990). Therefore, under most production conditions, cotton and corn would not be grown on soils that would inhibit the rate of degradation compared to what is seen at near neutral pH. On the other hand, potato may be grown at soil pH levels that approach those in which a substantial reduction in degradation rates has been shown to occur. However, effects of pH on degradation rates in the range of pH 5.0-5.8, which overlaps with potato growing conditions, has not been explored.

Studies have shown a substantial degradation (loss of biological activity) occurring rapidly in the first several weeks, with much slower subsequent breakdown (Tapp and Stotzky 1998, Palm et al. 1994, Palm et al. 1996). These experiments suggest that testing for persistence in the field should be determined over sufficiently long periods to assure an accurate assessment of degradation.

Many of the experiments examining persistence of Cry proteins reported in the published literature have apparently been conducted in bulk soils or soil components. Bulk soil generally does not support populations of microorganisms as high as in the rhizosphere or where plant residues are incorporated into the soil. Therefore, degradation rates under field conditions may be higher than those shown in bulk soil experiments.

It is important to consider that a number of factors are expected to influence persistence under actual field conditions, including: humic acid and clay content of the soil, clay type, pH, moisture, soluble ion content and type, and temperature, all of which affect microbial activity, composition, and population levels.

These factors may also may affect binding affinity of Cry proteins for soil components. Since these factors may vary considerably in the field, persistence of Cry proteins could likewise vary considerably. However, the conditions examined by the registrants generally replicate common field soil conditions, although performed in a laboratory setting. Consideration should be given to performing field tests of degradation under a range of conditions that may be expected under actual Bt crop cultivation.

A previously unconsidered issue regarding exposure of soil organisms to Cry proteins concerns the possibility that one of the Bt crops, specifically Bt 11 corn, exudes Cry1Ab protein into the soil (Saxena et al. 1999).

Exudation would likely cause continuous exposure of soil organisms to Cry1Ab protein. This situation differs from previous risk assessment considerations, which examined the effects of a single incorporation of Bt plant material or Cry protein, as would occur with incorporation at the end of the growing season. It is also possible that soil organisms could be exposed to higher levels of Cry1Ab than would occur with a single incorporation. Finally, since only Bt11 corn was examined, it is unknown whether the proposed exudation could occur in other Bt crops.

In the March 12, 2001 SAP Report No. 2000-07 on Bt Plant-Pesticides Risk and Benefit Assessment, the October 2000 Scientific Advisory Panel (SAP) concluded that published data at that time did not adequately address the persistence of Cry proteins from Bt crops in the soil. Since it is difficult to correlate the relevance of the published laboratory studies to field situations, the SAP recommended field studies be conducted in established Bt fields in a variety of soil types and climatic conditions.

The SAP suggested that the amount, accumulation and persistence of biological activity of Cry proteins in the soil are areas that should be investigated. However, the SAP also concluded that this data was not necessary for an EPA preliminary risk assessment but may be needed for a final assessment. In general, the Panel believed that studies on the mechanism Cry proteins enter soil (e.g., secretion, shedding of root hairs, degradation of biomass pollen) were primarily of academic interest. Knowledge of the potential environmental impacts is the important issue.

Exposure of Non-Target Organisms to Cry Proteins in the Soil

Most available data do not suggest toxic effects on non-lepidopterans from Cry1Ab or Cry1Ac proteins, or of non-coleopterans by Cry3A, even at doses from about 102 to 104 fold higher than the amounts estimated from the Saxena et al. paper.

The Bt11 Cry1Ab truncated protein was tested by the registrant on two species of Collembola (Folsomia candida and Xenylla grisea) and one species of earthworm (14 days) at 200 µg Cry1Ab/g feed with no observable effects. A published longer term continuous feeding study using transgenic cotton containing either Cry1Ab, Cry1Ac, or potato containing Cry3A at an estimated concentration of 10-20 PPM were performed on the collembolan

Folsomia candida and the orbatid soil mite Oppia nitens (Yu et al. 1997) for 7-8 weeks, and offspring examined for developmental effects. No adverse effects were detected in these studies for any of the three Cry proteins.

Therefore, high dose and continuous feeding studies on several invertebrates, including several important soil species, do not indicate likely adverse effects on non-lepidopteran species in the field.

The route of exposure of non-target organisms must also be considered. In the case of significant root exudation of Cry protein, some differences in exposure compared to single incorporation of transgenic plant material may be expected. Organisms that pass soil through their digestive systems, such as earthworms, could be exposed to higher levels of Cry protein due to exudation compared to a single incorporation of plant material. Organisms at higher trophic levels that feed on soil feeders may, secondarily, be exposed to toxin. Testing on birds and rodents however, showed that there was no toxicity to these species.

It is very difficult to determine the importance of shifts in the structure of microbial soil populations unless these changes can be associated with measurable ecological parameters. In most cases, such research has not been performed for microbial populations. Limited data do not indicate that Cry proteins have any measurable effect on microbial populations in the soil. No observable effects on microbial populations were found due to Cry1Ab or Cry1Ac protein exposure for up to 28 days for the former and 56 days for the latter in a single incorporation of 50 ng/g soil (Donegan et al. 1995). Small transient changes in microbial populations have been associated with some Bt transgenic plant material (cotton) rather than the transgenic Cry protein itself in experiments where transgenic plant tissue was used (Donegan et al. 1995).

The authors of the study suggested that the process of making the transgenic plant, e.g tissue culture which may introduce genotypic changes unrelated to the transgene, may have been responsible for these temporary population shifts. Other work comparing transgenic Cry3A potato and the non-transgenic isoline found no significant difference on the populations of several groups of soil microorganisms including fungal species diversity and three plant pathogenic fungi in this season long field study (Donegan et al. 1996).

Exposure of Soil Organisms to Cry Proteins from In Situ Roots or Incorporated Plant Tissue

Primary exposure to soil organisms has been considered to be from incorporation of crop residues at the end of the growing season, or to a lesser extent from deposition of pollen onto the soil. Therefore, degradation and possible accumulation of Cry proteins has been examined by determining degradation rates of Cry proteins, either in isolation or as expressed in the plant tissue, incorporated at a single point in time.

In addition, determination of degradation rates of Cry proteins in soil is more feasible using a discreet starting point. Estimates of total Cry protein incorporated into the soil have been based upon the biomass of total plant tissue, although it is not clear whether root biomass has been included in these calculations.

In contrast, the roots of a crop may be comparable in biomass to the above ground portions of the crop at the end of the growing season. During the growing season, soil organisms are exposed to roots and their contents. In particular, organisms that feed on living roots will ingest expressed Cry protein directly from this source.

Data for expression levels of Cry proteins in the roots are not available for all registered transgenic crops. Cry3A is expressed in potato tubers at 1.01 μg/g tissue (NAS/NRC 2000). An average of 20.2 μg/g total root protein (seasonal average for actively growing corn) for Bt11 corn containing Cry1Ab is expressed (Rusch and Kendall 1995).

Because of the variability of expression levels between transformation events, the levels of root expression in the other transgenic Bt crops cannot be predicted from these numbers or from expression in other tissues in those plants.

Similar to the case with above ground plant tissue, organisms that feed on root feeding organisms may be exposed indirectly and over an extended period of time. In addition, a significant amount of root tissue has been estimated to be lost during plant growth.

Estimates of loss of root tissue range from about 11-72 percent of total root tissue and about 4-20 percent for rhizodeposition of insoluble root material into soil (Newman 1985). This dead tissue may consist of exuded high molecular weight materials such as root cap mucilage, root cortical cells, or whole root tissue (Newman 1985).

The composition of this material is generally believed to consist largely of higher molecular weight materials, such as the structural components of the roots. It is therefore difficult to estimate whether the proportion of Cry protein in this material differs from that of living roots. Deposition of some Cry protein in the soil will likely occur during degradation of root tissue, in addition to Cry protein incorporated into the soil from the above ground parts of the crop at the end of the growing season.

As with the above ground portions of the plant, the root biomass increases during the growing season. Therefore, assuming that other factors are comparable, exposure of soil and soil organisms will be minimal early in the growing season and will increase with root volume. Therefore, the assumptions underlying a discrete, relatively short duration of exposure of both soil and soil organisms is likely to be inaccurate.

Several soil invertebrates have been used to examine the toxicity of Cry proteins. Submitted studies have examined high dose acute toxicity to earthworms and collembola, while published studies have examined the effects of longer term exposure to collembola, orbatid mites (Yu et al. 1997) and microorganisms (Donegan et al. 1995, Donegan et al. 1996). Also discussed above is new unpublished data (Stotzky 2000b) that indicates no apparent impact on a range of soil microorganisms, total biomass, etc. Results of all these studies have revealed little or no adverse effects.

The adequacy of these studies should be considered in the context of exposure during the growing season to root material, as well as incorporation of above ground material and roots at the end of the growing season. In particular, studies with all Cry proteins currently found in transgenic crops, and examining adults and offspring of the collembolan Folsomia candida and the orbatid mite Oppia nitens, both detritus feeders, showed no adverse effects at expression levels at or above those found in the above ground parts of the plant (Yu et al. 1997).

An additional study conducted by Saxena and Stotzky (2001) reported that Bt Cry protein released from root exudates and biomass of Bt corn has no apparent effect on earthworms, nematodes, protozoa, algae, bacteria, actinomyces and fungi in soil in spite of the fact that enough detectable Cry protein is bound to soil particles to show toxicity to the target pest. In addition, toxicity studies of above ground invertebrates, discussed elsewhere, also show few adverse effects to taxa that are not closely related to those known to be affected. While not soil organisms themselves, these data generally suggest the lack of substantial adverse effects to soil invertebrates.

Available studies on the impact of transgenic Cry producing plants indicates that adverse effects on soil microorganisms are unlikely. No effects have been seen due to the protein itself, and only a minimal, transient increase observed in soil microbes attributed to the transgenic cotton plant tissue rather than the Cry protein expressed in that tissue (Donegan et al. 1995). No adverse effects have been observed in a similar season long field study with Cry3A potato (Donegan et al. 1996).

Horizontal Transfer of Trangenes From Bt Crops to Soil Organisms

Concern has been expressed about the possible transfer of transgenes from crop plants to related or unrelated species of plants or other organisms. Also of concern is the possibility of horizontal gene transfer, that is transfer to unrelated species. Horizontal gene transfer, regardless of the source of the gene, has been considered in the literature, and evidence presented for possible impact on the evolution of organisms over long periods of time (Smith et al. 1992).

However, the likelihood of such transfer over a human time scale, and where the transgene has a positive impact on the fitness of the recipient, has not been demonstrated. Horizontal transfer of several traits might be of concern for the current Bt crops, for example the cry genes and antibiotic resistance genes.

There is no evidence that horizontal gene transfer occurs from plants to microbes or bacteria to bacteria. Bt genes which naturally occur in many soils have never demonstrated horizontal gene transfer. The October 2000 SAP concluded that the "horizontal gene transfer assessment appear to be adequate, and no additional data are probably necessary for the Agency to complete a risk assessment."

Determination of Cry Protein Levels in Soils Following Several Years of Bt PIPs

In the March 12, 2001 SAP Report No. 2000-07 on Bt Plant-Pesticides Risk and Benefit Assessment, the October 2000 Scientific Advisory Panel (SAP) concluded that published data at that time did not adequately address the persistence of Cry proteins from Bt crops in the soil. Since it is difficult to correlate the relevance of the published laboratory studies to field situations, the SAP recommended field studies be conducted in established Bt fields in a variety of soil types and climatic conditions.

The SAP suggested the determination of the amount, concentration, and persistence of biological activity of Cry proteins in the soil are areas that should be investigated. The EPA agrees with the SAP that actual field data on Cry protein levels in soil will yield relevant data on persistence and natural variation of plant-produced Bt proteins in soil. If high levels of Cry proteins are found in field soils, reevaluation of the risks to certain non-target organisms might be required. Therefore, EPA is requiring additional supplementary studies regarding Cry protein in soil.

The Agency is requiring testing of Cry1Ab and Cry1F protein under a range of conditions typical of Bt corn cultivation. EPA requires each registrant or the registrants in cooperation to submit test protocols before the studies are actually conducted. In general, the Agency anticipates that soils would be sampled from fields where Bt corn has been grown continuously for at least 3 years compared with fields where no Bt crop has been grown. These paired fields would include several locations throughout the corn growing area of the US representing different soil and climatic variations. The Agency anticipates that samples would need to be taken 2 or 3 times during the growing season.

INSECT RESISTANCE MANAGEMENT IN BT CROPS

Insects, fungi, and weeds developing resistance to pesticides are well documented in agriculture. As resistance begins to develop, more pesticide is needed to achieve control until total failure of that pesticide occurs.

Insect resistance management (IRM) is the term used to describe practices aimed at reducing the potential for insect pests to become resistant to a pesticide. Academic and government scientists, public interest groups, and organic and other farmers have expressed concern that the widespread planting of genetically transformed plants will hasten the development of resistance to Bacillus thuringiensis (Bt) endotoxins. Sound IRM will prolong the life of Bt pesticides and adherence to the plans is to the advantage of growers, producers, researchers, and the American public.

EPA's strategy to address insect resistance to Bt is twofold:

(1) mitigate any significant potential for pest resistance development in the field by instituting IRM plans, and

(2) better understand the mechanisms behind pest resistance.

IRM is important for transgenic crops expressing Bacillus thuringiensis (Bt) insecticidal proteins (Bt crops) because insect resistance poses a threat to future use of microbial Bt pesticides and Bt technology as a whole.

The goal of IRM is to have the target pest continue to be susceptible to the pesticide. Each IRM program consists of strategies to reduce the likelihood that insect resistance will develop and strategies to manage insect resistance once it occurs. Bt IRM is of great importance because of the threat insect resistance poses to the future use of Bt microbial pesticides and Bt plant-incorporated protectants.

Specific IRM strategies, such as the high dose/structured refuge strategy, will mitigate insect resistance to specific Bt proteins produced in corn, cotton, and potatoes. Effective insect resistance management can reduce the risk of resistance development.

At a Federal Insecticide, Fungicide, and Rodenticide Act (FIFRA) Scientific Advisory Panel (SAP)1 meeting in March of 1995, EPA laid out a multi-faceted program that the agency considered appropriate for Bt crop products.

The elements are: knowledge of pest biology and ecology, dose (level of toxin expressed in the Bt crop), refuge design and deployment (non-Bt plants producing Bt-susceptible insects), cross-resistance between different Bt proteins, effective field monitoring for insect resistance, remedial action if resistance occurs, integrated pest management, development of alternate modes of action, and grower education.

SCIENTIFIC BASIS FOR AN IRM PROGRAM

As an IRM program is developed and implemented, each pest's unique biology must be factored into the plan. For example, how far do the larvae move within the field and how far do the adults move affects the distance between the refuge and the Bt crop.

The susceptible insects from the non-Bt refuge need to be in close enough proximity to randomly mate with the resistant insects that emerge from the Bt fields to produce heterozygous offspring that are fully susceptible to the Bt protein.

Additional important questions that need to be addressed are how many generations of insects are produced each year, what is the mating behavior and the oviposition behavior, what is the host range of the insect, population dynamics, pest ecology, and if possible, the genetics and mechanism of resistance and the frequency of resistance alleles in the insect population.

In addition, how the crop is grown including other pest management practices, when it matures, the extent of the acreage, and the overlap in distribution with other Bt crops are all important in the development of an appropriate program.

Current resistance management programs for Bt plant-incorporated protectants (PIPs) are based on the use of both a high Bt toxin concentration coupled to the use of structured refuges planted nearby to provide sufficient numbers of susceptible adult insects.

The "high dose/structured refuge strategy" assumes that resistance to Bt is recessive and is conferred by a single locus with two alleles resulting in three genotypes: susceptible homozygotes (SS), heterozygotes (RS), and resistant homozygotes (RR).

It also assumes that there will be a low initial resistance allele frequency and that there will be extensive random mating between resistant and susceptible adults. Under ideal circumstances, only rare RR individuals will survive a high dose produced by the Bt crop. Both SS and RS individuals will be susceptible to the Bt toxin.

A structured refuge sets aside some percentage of the crop land for non-Bt varieties of that crop. The refuge provides for the production of susceptible (SS) insects that may randomly mate with rare resistant (RR) insects surviving the Bt crop to produce susceptible RS heterozygotes that will be killed by the Bt crop. This will remove resistant (R) alleles from the insect populations and delay the evolution of resistance. The 1998 SAP defined a high dose as 25 times the toxin concentration needed to kill susceptible larvae. This standard has been used by EPA since 1998.

To be effective, an IRM plan must be specific to the target pest, the crop, and to the class of pesticide being used. The use of a high dose strategy is often not the best approach for chemical pesticides sprayed on a crop and pesticide rotation is one of the preferred approaches. Annually rotating between Bt crops and non-Bt crops treated with conventional insecticides is likely to wipe away many of the environmental benefits of increased non-target organisms we are seeing in areas where Bt crops are frequently grown.

SPECIFIC REFUGE REQUIREMENTS

The Agency has determined that a 20% non-Bt field corn refuge requirements for Bt field corn grown in the Corn-Belt and a 50% non-Bt corn field refuge requirements for Bt field corn grown in cotton-growing areas is scientifically-sound, protective, feasible, sustainable, and practical to growers. EPA believes that the use of predictive models provides confidence that resistance will not evolve to any of the target pests (i.e., European corn borer, corn earworm, southwestern corn borer, fall armyworm, and other stalk-boring pests) under the time frame of the registrations.

For Bt sweet corn, no specific refuge requirements are necessary because sweet corn is typically harvested much earlier than field corn, 18-21 days after silking, and before most lepidopteran larvae complete development.

However, to mitigate the development of resistance, EPA has determined that crop residue destruction is necessary within 30 days. This practice will likely destroy any live larvae left in Bt sweet corn stalks and prevent overwintering of any resistant insects. At this time, the Agency believes that available empirical data substantiate the success of the 5% external unsprayed, 20% external sprayed, and 5% embedded structured refuge options to delay insect resistance to Bt cotton.

However, EPA believes that it is imprudent to allow the 5% external, unsprayed refuge option for more than a limited period of time because current data indicates that this option has a significantly greater likelihood of insect resistance than either of the other refuge options.

The 2000 SAP stated that the external, unsprayed option poses the highest risk to resistance evolution especially for cotton bollworm. Because of the greater risk of resistance development, the external, unsprayed option expires after three growing seasons (September 30, 2004).

Presently, the Agency is mandating additional improvements to the current Bt corn (field and sweet) and Bt cotton IRM programs that will require:

(1) anyone purchasing Bt corn and Bt cotton to sign a grower agreement which contractually binds the grower to comply with the IRM program and that there will be a mechanism by the year 2003 by which every grower affirms their contractual obligations to comply with the IRM program,

(2) an ongoing IRM education program,

(3) an ongoing IRM compliance monitoring program including a third party compliance survey and mechanisms to address non-compliance,

(4) and ongoing insect resistance monitoring program for each target insect pest,

(5) remedial action plans to be implemented if resistance does develop, and

(6) annual reporting of the IRM (and other) activities. No other pesticide products than the Bt crop products have such extensive IRM requirements.

In considering the new IRM regulatory program for Bt plant-incorporated protectants (PIPs), EPA considered not only the science, but also factors such as grower costs and compliance, resistance monitoring, and remedial action if resistance should occur.

EPA considered four important areas: Farmer Actions, Compliance Monitoring, Insect Resistance Monitoring, and Remedial Action Plans. Each of these areas was considered in making the decision to continue the registrations of the Bt PIPs.

FARMER ACTIONS

Farmer adoption of IRM requirements is critical to the long-term, sustainability of IRM strategies for Bt crops. The first essential farmer action is to become familiar with the IRM requirements for his (or her) Bt crop. Without farmer implementation of appropriate IRM strategies then pest resistance cannot be mitigated. Each farmer must sign a contract or grower agreement indicating that the farmer will abide by the IRM requirements and the farmer receives a technical bulletin describing the latest requirements.

Educating the farmer also includes making sure that the farmer is aware of any changes that have occurred since he/she last grew the Bt crop, whether it was last year or two or more years ago.

Educational materials are provided by each company and education sessions are held by the companies, sometimes by the seed dealer, the commodity group, and often by the United States Department of Agriculture's Cooperative Extension Service. Information is also provided through the internet and through newsletters and other media.

Once educated to the requirements, it is the farmer's responsibility to plant and manage the refuge. It must be the correct size (such as 20 percent non-Bt corn to 80 percent Bt corn in the Corn Belt), it must be placed at the correct distance so that any Bt-resistant insects coming from the Bt crop will easily find mates from the Bt-susceptible insects coming from the refuge, and the farmer must plant a refuge using a crop variety compatible to the Bt crop in the time that adults would emerge from the refuge and the Bt crop.

Farmers also play an important role in supplementing monitoring by reporting any failure of the Bt crop to control the target pest. Because farmers pay an extra fee when they buy the Bt seeds, they have an incentive to complain to the company who sold them the seed if it is not preforming correctly. An important farmer action in the overall IRM program, is that the farmer cooperate and accurately respond to questionnaires and surveys on what activities the farmer has actually done.

COMPLIANCE MONITORING PROGRAM

EPA recognizes that compliance is a complex issue for Bt crops and IRM; therefore, a balance must be achieved between refuge size and deployment with grower compliance. Currently, the financial burden of implementing refuge requirements is borne primarily by the growers. Increasing refuge size and/or limiting refuge deployment to better mitigate the risk of resistance will likely increase costs to growers and result in a higher rate of grower non-compliance.

The Bt crop reassessment has greatly strengthened the compliance monitoring to increase the likelihood of IRM adoption, to measure the level of compliance, and to institute penalties should non-compliance become a significant problem.

Until recently, monitoring for farmer compliance to the IRM program has been largely voluntary, but now it is mandatory. Key to this program is the grower agreement between the company and the farmer indicating the farmer will abide by the IRM requirements. Although it varies a bit by crop, the program is basically a tiered approach of actions depending on the results of a grower survey.

The survey is conducted by a third party, independent organization using funds provided by the companies. The survey questions are developed in consultation with academic and government researchers knowledgeable on the subject and there is an EPA review of the survey also.

The survey focuses on areas of highest risk, typically those areas of highest adoption. If the survey indicates that an area of the country is not fully complying with the requirements, increased education and more intense surveying will be implemented in that area.

The degree of increased effort may be directly related to the type of problem. For example, if bad weather conditions cause farmers to plant their refuge late, that is quite different than if farmers in an area decide it is unimportant to plant the refuge at all.

In addition, typical on-farm visits conducted by the companies and/or their representatives will report on farmers who are or are not complying with the requirements and what follow-up actions need to be taken. Any farmer determined to be out of compliance, will automatically receive an on-farm inspection the following year.

If that farmer is still found to be significantly out of compliance, that grower would be denied the use of the Bt crop the following year.

While that farmer may be able to buy the technology the third year, that farmer would again automatically receive an on-farm visit during that growing season and if that farmer again is out of compliance, the farmer would be denied the use of the technology permanently.

INSECT RESISTANCE MONITORING

Monitoring has been part of the requirements of the Bt crop products registration from the beginning.

The goal is to detect insect resistance before it occurs in the field or before it spreads and if possible, prevent the development of resistance by detecting increased pest susceptibility. EPA's program includes monitoring for the important target pests.

To be effective, the plan requires sensitive tools to be in place that can detect changes in resistance allele frequency to the particular Bt protein and the ability to differentiate between a natural variation in the population and a trend indicating that resistance is likely to happen soon or may have already happened.

In one of the early steps in developing this program, EPA established a working definition for what is resistance versus what is natural tolerance variation and our analysis was reviewed by our Scientific Advisory Panel for confirmation.

In addition, EPA, working with the pesticide companies, has established definitions for suspected versus confirmed resistance. An additional consideration is the time required to "confirm" resistance. The basic resistance monitoring program is to gather the target insects in an adequate sample size from an appropriate number of locations and to test for susceptibility to the Bt protein. Samples can be collected from various live stages.

Adults might be collected from light or pheromone traps that attract the moths or larvae and/or eggs masses might be collected either from Bt fields or other crop or non-crop areas. Depending on the life stage collected, the insects might have to be reared to a stage when they can be fed on the appropriate Bt protein to determine their level susceptibility to the insect toxin.

Resistance monitoring is a difficult and imprecise task. The chances of finding a resistant larvae in a Bt crop depend on the level of pest pressure, the frequency of resistant individuals, the location and number of samples that are collected, and the sensitivity of the detection technique.

Therefore, as the frequency of resistant individuals in the insect population increases or the number of collected samples increases, the likelihood of locating a resistant individual increases. The likelihood of resistance is dependent of the genetics and mechanism of resistance for a particular pest.

A resistance monitoring program is more important when the predicted time to resistance is small rather than when the models predict that resistance is expected to be delayed for a very long time.

Based on predictive models, level of adoption, and compliance for European corn borer resistance to Bt proteins expressed in field corn predict 75 years or more until resistance would develop, but for cotton bollworm, tobacco budworm, and pink bollworm, the predicted years to resistance to Bt proteins expressed in cotton is much shorter.

The resistance monitoring program needs to consider the pest biology and ecology, population dynamics, genetics of resistance, mechanism of resistance, sampling methodology, bioassay methodology, standardization procedures, detection technique and sensitivity, and the statistical analysis of the probability of detecting resistance.

To determine if refuges or any other resistance management tactics are working, one must track the frequency of resistance in field populations. With typical bioassays used for resistance monitoring, resistance cannot be detected readily, when the allele is recessive (as often is the case) and rare.

For example, if the frequency of a recessive resistance allele is 0.001, only one in a million individuals is expected to be a resistant homozygote (carrying two resistance alleles) capable of surviving exposure to a high concentration of the Bt protein.

There are a number of issues associated with this program. The first issue is sample size.

The number of samples and number of locations that need to be sampled is dependent on the pest biology and ecology and population dynamics. If the genetic variation in an insect is known, then sampling strategies can be constructed with a greater probability of detection and a low probability of non-detection. Both factors must be considered to reduce the likelihood of both Type 1 (false positive) and Type 2 (false negative) errors.

Sampling should also be done in a uniform fashion. Uniformity and standardization in the bioassays is also critical to the interpretation of monitoring information. Related to sample size is finding enough insects to test.

Sampling insects exposed to the Bt crop is preferred, but if sampling is predominately in the Bt crop, then there are few, if any, larvae of the target insect to be found in most Bt fields. This means that sampling methods need to be adapted to either collect adults or egg masses to generate the volume of individuals that are needed to increase the probability of detecting resistance or samples are taken from non-Bt fields.

Current resistance monitoring plans in the United States have a goal to collect at least 250 individuals from any one location with a target of least 20 locations for tobacco budworm and cotton bollworm, pink bollworm, and European corn borer. Additional sampling for southwestern corn borer is focused in those areas of the Corn Belt in which this pest is an economic problem. The greater the number of samples and number of locations, the greater the probability that resistant individuals will be collected.

Another issue is the sensitivity of the detection methods. If resistance is recessive (rather than dominant or co-dominant), resistance is less likely to develop, but it's more difficult to detect. It is useful to know the frequency of the resistance allele in the natural population. Estimates of the frequency of resistance alleles have been determined based on laboratory-selection experiments (surrogates for what might happen, but not necessarily what will happen in the field).

Field verification of resistance allele frequency require reliable and sensitive detection methods. However, if extremely sensitive detection methods, especially if resistance is recessive, are available and economically feasible to use, changes in resistance allele frequency (and verification of estimates) prior to any signs of field failure and create opportunities for proactive, adaptive IRM.

Additional tests include grower reports of unexpected damage, sentinel plots or use of in-field screening procedures, screening against resistant test stocks (allelic recovery method), and use of in-field detection (using DNA markers) kits.

Gould et al. (1997) used a series of genetic crosses with test stocks of highly resistant tobacco budworm (YHD2) selected on Cry1Ac in the laboratory to estimate the resistance allele frequency in a natural population of tobacco budworm.

This method can identify recessive or incompletely dominant resistance alleles from field-collected males. By using an assay that discriminates between heterozygotes, they could establish which wild males carried a resistance allele.

Using this allelic recovery method, Gould et al. (1997) estimated the resistance allele frequency to be 1.5 X 10-3. This method is only useful when there are previously identified resistance alleles. Gahan et al. (2001) were able to identify the mechanism of resistance in this YHD2 line and be the first develop a DNA-marker that might be used in the field to screen for resistance.

In addition to sampling and detection sensitivity, other equally complex issues are related to cost and feasibility. It would be virtually impossible and economically prohibitive to sample every farm in which Bt crops are used.

For example, there are approximately 14,000 Bt cotton producers (out of approximately 25,000 cotton producers). These producers planted approximately 4.5 million acres of Bt cotton in the 2000 growing season. Current resistance monitoring programs have focused sampling in areas of highest adoption of the Bt crops as the areas in which resistance risk is greatest. There were approximately 20 million acres of Bt corn planted in the 2000 growing season.

The cost of the US monitoring program is borne chiefly by the companies although academic institutions and the U.S. Department of Agriculture researchers who carry out the bioassays probably bear some costs (i.e., University of Nebraska for European corn borer, University of Arizona for pink bollworm, University of Missouri for southwestern corn borer, and USDA/ARS at Stoneville, Mississippi for tobacco budworm and cotton bollworm).

REMEDIAL ACTION PLAN

EPA requires a remedial action plan be available in the unfortunate situation that resistance is suspected or actually does develop. Again, as for resistance monitoring plans, remedial action plans are specific for the crop and pest.

For example, because the pink bollworm is predominately a pest of cotton in the western US and has such a different biology than the other two target pests of Bt cotton, the remedial action plan for pink bollworm is quite different than for cotton bollworm and tobacco budworm in the southeastern US. These plans define not only suspected and confirmed resistance, but the key steps and actions needed if and when resistance develops.

Generally, if resistance is confirmed, the farmers involved will treat their Bt crop with alternative pest control measures. This might be a chemical pesticide known to be highly effective against the insect or it might mean measures such as crop destruction. In addition, the sales and distribution of the Bt crop would be suspended in that area and the surrounding area until it can be determined that insects in that area have regained their susceptibility to the Bt protein.

There would also need to be increased monitoring to define the remedial action area(s). Other remedial action strategies include increasing refuge size, changing dispersal properties, use of sterile of insects, or use of other modes of actions. Geospatial surveys would help define the scale of remedial action and where to intensify monitoring.

Because no field resistance has yet been found to any of the Bt crops, all of these tactics are untested. However, EPA believes that a key attribute of these plans are having involvement in the plan's development by the local farmers who would be affected most by the loss of this technology.

So far there is only a regional remedial action plan for the Arizona area where pink bollworm is the chief pest controlled by Bt cotton. An interim remedial action plan is required and is being revised to address tobacco budworm and cotton bollworm resistance to Bt cotton, and key economic pests of cotton in the mid-South and Southeastern US. There is also a general remedial action plan to address resistance to European corn borer, southwestern corn borer, and corn earworm.

CONCLUSION

The four IRM activities (farmer actions, compliance monitoring, insect resistance monitoring, and remedial action plans) need to be balanced. To some extent, they are at least partial substitutes for each other.

In other words, if the refuge is extremely large (95%), there is virtually no need to monitor for insect resistance because resistance is so unlikely to occur.

However, having a 95 percent refuge would eliminate many of the benefits to growers and the environment for cotton growers.

Monitoring every Bt field for insect resistance reduces the need for a compliance program, but such an intensive effort is infeasible and extremely costly.

In our regulation of these Bt products, EPA has attempted to balance these activities. EPA believes that the increased quality and substance of the compliance monitoring and resistance monitoring programs required through our completed reassessment can compensate to some extent for the small refuge size for Bt cotton.

In addition, EPA has required additional data on the effect of alternate plant hosts and alternative modes of actions on delaying cotton bollworm resistance to Bt cotton. EPA believes that technological improvements to detect resistance earlier in the field will result in scientifically valid methods that will be cost-effective for insect resistance management in the future. Our faith in improvements ahead comes from knowing that academic, company, and government research continues to be strong in the area of IRM.

BIBLIOGRAPHY

Bainbridge, W.S. : *Societal Implications of Nanoscience and Nanotechnology*, Kluwer, Dordrecht, 2001.

Berry, Brian L.: *The Social Burdens of Environmental Pollution: A Comparative Metropolitan Data Source*, Cambridge, Ballinger Publishing Co., 1977.

Daubenmire, R. F.: *Plants and Environment*, New York, Wiley, 1947.

Etzkowitz, H.: *Nano-Science and Society: Finding a Social Basis for Science Policy*, Dordrecht, Kluwer, 2001.

Faden, R. R., and T. L. Beauchamp: *A History and Theory of Informed Consent*, New York, Oxford University Press, 1986.

Gelobter, Michael: *The Distribution of Air Pollution by Income and Race*, Berkeley, University of California, 1989.

Goodnick, M.: *Transport in Nanostructures*, Cambridge, Cambridge University Press, 1997.

Hayles, N.K. : *Nanoculture: Implications of the New Technoscience*, Bristol, Intellect Books, 2004.

Hofrichter, Richard: *The Theory and Practice of Environmental Justice*, Philadelphia, New Society Publishers, 1993.

Johnson, Rebecca L.: *The Greenhouse Effect: Life on a Warmer Planet*, Minn., Lerner Publications, 1993.

Kane, R. A., and A. L. Caplan: *Everyday Ethics: Resolving Dilemmas in Nursing Homes*, New York, Springer Publishing, 1989.

Kruvant, W.J.: *People, Energy, and Pollution*, Cambridge, Ballinger Publishing, 1975.

Moravec, Hans: *Mind Children: The Future of Robot and Human Intelligence*, Cambridge, Harvard University Press, 1988.

Norio Taniguchi: *Nanotechnology: Integrated Processing Systems for Ultra-precision and Ultra-fine Products,* New York, Oxford University Press, 1996.

Postman, Neil: *Technopoly: The Surrender of Culture to Technology,* New York, Vintage Books, 1992.

Roobeek, A: *The Biotechnology Revolution?*, Oxford, Blackwell, 1995.

Shew, A.: *Discovering the Nanoscale,* Amsterdam: IOS Press, 2004.

Sontheimer, Sally: *Women and the Environment: Crisis and Development in the Third World,* New York, Monthly Review Press, 1991.

Stephen M. Gooanick: *Transport in Nanostructures,* Cambriage, Cambridge University Press, 1997.

Szasz, Andrew: *Ecopopulism: Toxic Waste and the Movement for Environmental Justice,* Minneapolis, University of Minnesota Press, 1994.

Tenner, E.: *Nanotechnology and Unintended Consequences,* , Dordrecht, Kluwer, 2001.

Veatch, R. M.: *The Patient as Partner: A Theory of Human-Experimentation Ethics,* Bloomington, Ind.: Indiana University Press, 1987.

Vincenzo, M. Venturi, and A. Credi: *Molecular Devices and Machines: a Journey into the Nano World,* Weinheim, Wiley-VCH, 2003.

Zupan, J.M.: *The Distribution of Air Quality in the New York Region,* Baltimore, Johns Hopkins University Press, 1973.

INDEX

□□□